Medical Technology Series

CYTOGENETICS

Jean H. Priest, M.D.

Assistant Professor of Pathology and Pediatrics
University of Colorado Medical Center
Denver, Colorado

Lea & Febiger   **1969**   **Philadelphia**

SBN-8121-0193-6

Published in Great Britain by

Henry Kimpton, London

Library of Congress Catalog Card Number 76-78542

Printed in the United States of America

Other titles available

Foreword

The science of human genetics is advancing rapidly as knowledge and application of basic principles increase. The study of genetics is no longer limited to a few specialized scientists but has become a common meeting ground for many with diversified training and interests. Cytogenetics is the study of chromosomes and genes, the cellular constituents concerned with heredity. Almost all who become knowledgeable in this field develop an avid curiosity about mechanisms of abnormalities and methods to determine them.

The purpose of this monograph is to guide the performance of certain techniques in laboratories and to teach these techniques to those wishing to enter this challenging field. Basic knowledge essential for the interpretation of cytogenetic studies, outlines of techniques, and references are presented. Included are an introduction to the study of cytogenetics and human chromosome abnormalities, nomenclature, methods for performing chromosome and sex chromatin studies, microscope techniques, and principles of tissue culture.

Only through the development of chromosome methods has a new knowledge of birth defects and the mechanisms of chromosomal genetics evolved. It is hoped that through the knowledge gained from referring to this monograph, readers will be able to cope better with the rapidly increasing number of requests for cytogenetic studies and will be stimulated to work toward improving methods, as well as applying theories to facts.

CATHERINE W. ANTHONY, M.D.
Chief, Laboratory Service
Veterans Administration Hospital
Denver, Colorado

Preface

This book is intended to assemble and correlate information related to human cytogenetic methodology. Because terminology plays an important role in the understanding of this area of study, two chapters are devoted to definitions. Cytogenetic nomenclature is discussed in Chapter 3, and tissue nomenclature in Chapter 13. In addition, the italicized words in Chapters 1 and 2 and in the introductions to the other chapters are defined in the Glossary at the end of the book.

In the Contents I have listed a subtitle if I considered it important for the reader to know that the subject is covered, even if just by references. Subjects for which references only are given are indicated by an asterisk (*).

I would like to thank Dr. Robert Priest for his help and encouragement during preparation of this manuscript. Dr. Robert Shikes and Miss Harriet McKelvey gave valuable suggestions during proofreading of the final drafts. Mrs. LaVonne King typed the manuscript with great care.

This book is dedicated to all those individuals who find chromosomes exciting.

JEAN H. PRIEST
Denver, Colorado

Contents

Medical Technology Series

1

An Introduction to Cytogenetics

MITOTIC CHROMOSOMES (Fig. 1-1)

In 1956 Tjio and Levan[15] first correctly described the normal human *diploid* chromosome number as 46. Using modifications of techniques previously developed by Hsu,[6] they examined mitoses from human embryonic lung cells. The two most important parts of these techniques for demonstrating mitotic chromosomes and facilitating the analysis of number and morphology, were (1) the use of colchicine to "harvest" the mitoses and (2) the use of hypotonic solution on the cells prior to fixation to spread out the chromosomes. The present classification of human mitotic (*somatic*) chromosomes at metaphase is based on the Denver Classification (1960), with revisions at the London Conference (1963) and the Chicago Conference (1966).[2] The chromosomes are serially numbered 1 to 22 as nearly as possible in descending order of length. The *sex chromosomes* continue to be referred to as X and Y. The *autosomes* are classified into seven groups, A to G (Figs. 1-2 and 1-3; also see Chapter 3).

As has already been implied, the examination of mitotic chromosomes depends on obtaining large numbers of dividing somatic cells. There are three main ways to accomplish this situation. Tissues already

2

dividing rapidly *in vivo,* such as bone marrow, may be examined without *in vitro* culture. This method has the advantage of speed. It is also free of artifacts possibly introduced by culture outside the living individual, but has the disadvantages that only certain tissues may be examined and in general the choice of metaphases for evaluation is not great. The second method, and the one used most frequently for the clinical screening of human chromosome defects, involves *short-term in vitro culture* of peripheral blood. Although the *mitotic index* (rate of cell division) is low in circulating nucleated blood cells, it has been shown that various *mitogenic* agents, including a bean plant extract, phytohemagglutinin, can induce lymphocytes to undergo division and *"blastoid" reaction* to a cell of less mature appearance.[12] The burst of mitosis (Fig. 1-4) is utilized to study human metaphases. This type of short-term culture has the advantages of speed, ease of tissue culture techniques, and ease of obtaining the specimen to be cultured. Disadvantages are variability in production of the mitogenic response and short life of the culture so that it cannot be maintained for further studies. The third way to obtain large numbers of dividing somatic cells is by *long-term culture* techniques. A *primary explant* or tissue specimen is obtained,

3

often by pinch skin biopsy from a living human subject or from any available and desirable tissue at surgery or autopsy. Cells that grow from the primary explant in general are *fibroblast-like* in appearance (Fig. 1-5), grow as a *monolayer* in tissue culture containers, and have a limited lifetime of about 50 *population doublings* in serial culture.[4, 5] Advantages to long-term culture techniques are the availability of

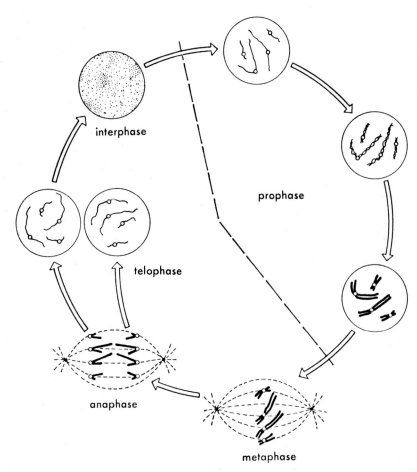

Figure 1-1. *Mitosis. Two pairs of chromosomes are shown. Only the nucleus is represented. Following anaphase, new nuclear membranes are formed, and the cell cleaves to give 2 daughter cells identical in genetic content to the parent cell. (From* An Introduction to Human Genetics *by H. Eldon Sutton. Copyright 1965 by Holt, Rinehart and Winston, Inc. Reproduced by permission of Holt, Rinehart and Winston, Inc.)*

4

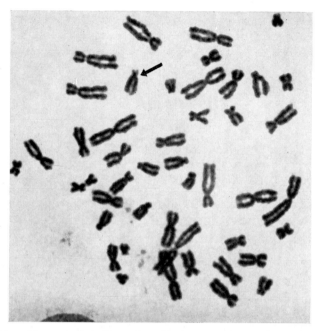

Figure 1-2. *Human* diploid *male mitotic metaphase chromosomes as seen through the 100X (oil immersion) objective of a light microscope. Giemsa stain. Each of the 46 mitotic chromosomes is visibly divided into 2* chromatids. *Coiled* chromatid *substructure can be seen. A* centromere *(primary constriction or kinetochore) is a nonstaining area at the joining of 2* chromatids. Centromere *positions are classified as median, submedian, and subterminal, forming* chromatid arms *of different lengths. Lighter or nonstaining areas* (secondary constrictions), *as well as darker staining areas are seen along the* chromatid arms. *Usually these areas vary from one preparation to another, although* secondary constrictions *may be characteristic of a particular chromosome.* Satellites *(arrow) are visible on the distal ends of the short* arms *of some chromosomes with subterminal* centromeres. *Chromosome distribution is random except for the tendency of the* sex chromosomes *to be peripheral in location.*

cells for further study and the opportunity to examine many kinds of tissues. Disadvantages are the time and effort involved in maintaining cells in serial culture. Furthermore, a single morphological cell type grows from the initial tissue and in this sense is not representative of the entire tissue sampled.

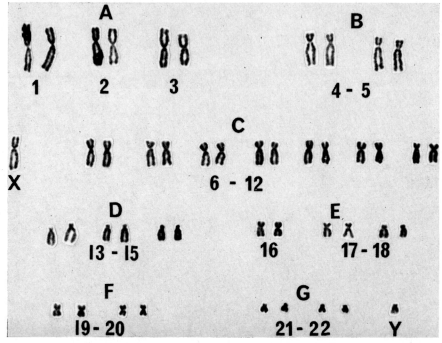

Figure 1-3. Karyotype *of the human* diploid *male metaphase chromosomes seen in Figure 1-2. The classification of human metaphase chromosomes is based on* centromere *position and relative* length. *The chromosomes are arranged in order of decreasing total* length. *There are 22* autosome *pairs and 1* sex chromosome *pair (XX in the female and XY in the male). The* autosome *pairs are divided into 7 groups (A to G). Within groups only pairs 1, 2, 3, 16, 17, and 18 can be distinguished with certainty. The Y chromosome resembles those in the G group and sometimes can be distinguished by larger size, less distinct* centromere, *poor definition of terminal regions of the long* arms, *and less divergence of the long* arms.

After somatic metaphases are obtained by the *in vitro* methods just described, they are accumulated by the addition of colchicine or other compounds known to arrest cells in metaphase. The length of application of colchicine is limited because in addition to stopping the cells in metaphase (theoretically all the cells could be accumulated in metaphase), if applied in too strong a concentration or for too long a time, it causes further condensation and distortion of the chromosomes so that they are no longer suitable for analysis according to the standard nomenclature. In addition, metaphase arrest is not complete or perma-

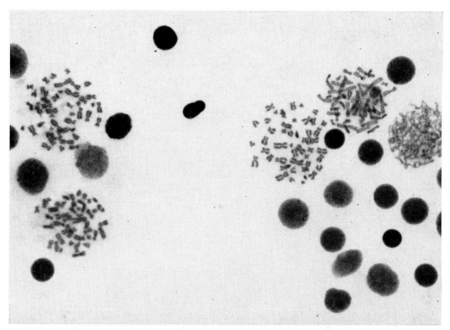

Figure 1-4. *Mitotic figures in a human peripheral blood culture, as seen through the 40X objective. Giemsa stain.*

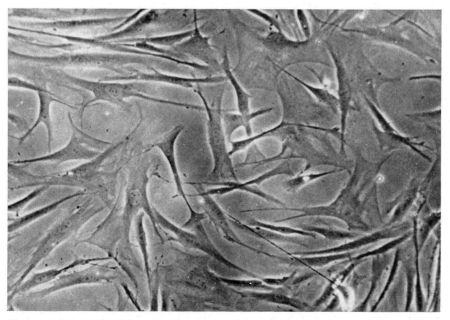

Figure 1-5. *Fibroblast-like human diploid cells growing in serial monolayer culture. Phase contrast microscopy.*

7

nent. In some cases long application of colchicine will cause *endoreduplication* (Fig. 1-6) and *polyploidy*, thus introducing chromosome errors not present in the original tissue.

The remaining procedures for chromosome analysis involve application of hypotonic solution to expand the cells, fixation, squash or air-drying techniques to flatten the cells on the slides, staining with a *DNA* stain, analysis and photography of metaphases through the oil immersion lens. The process of analysis with subsequent preparation of a *karyotype* or a photograph of chromosomes of a cell arranged according to

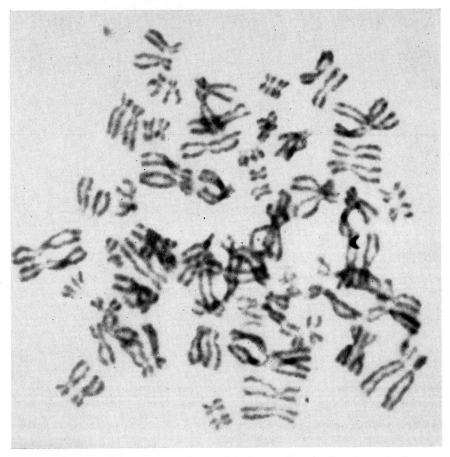

Figure 1-6. *A human metaphase showing* endoreduplication *of chromosomes.*

8

the standard classification is time consuming. Computer techniques are being formulated to shorten the time required for analysis of photographs of chromosomes.[8]

The standard nomenclature of human chromosomes already described applies to fixed material examined through the light microscope. The description of chromosomes is therefore static and subject to distortions of fixation as well as to limitations of resolution of the light microscope. Furthermore, metaphase DNA is condensed and biologically inactive, in the sense that no *DNA replication* or *RNA transcription* is taking place. Therefore, the question arises as to how we can proceed to more dynamic characterizations of chromosomes, keeping in mind that present classifications are very useful, particularly for the diagnosis of abnormalities in chromosome morphology resulting in congenital defects.

If *tritiated thymidine* is supplied to cells in culture, the labeled thymidine is incorporated specifically into *DNA* during the *synthesis period (S)* of the *cell life cycle,* and the label may be localized exactly over the chromosomes at next metaphase by *autoradiography*. Therefore, certain aspects of the *replication* behavior of the chromosomes during interphase can be recorded and studied at metaphase. There is asynchrony of DNA replication between *homologous* chromosomes and in different segments of single chromosomes. One X chromosome in XX individuals starts and finishes replication later than the other X chromosome and the autosomes. Because X chromosomes cannot be distinguished from other chromosomes of similar morphology according to the Denver classification, the unique replication behavior of one X is the only way it can be separated from the other chromosomes at metaphase.

Studies of chromosomes in the living state by means of phase contrast *cinemicroscopy* present an entirely different morphologic picture[1] which can be appreciated best by watching the phase movies. Electron microscopic observations of human metaphase chromosomes do not at present resolve the gap between the morphology seen by the light microscope and the *Watson-Crick model* for the structure of the *DNA* molecule. In the mammalian chromosome there is evidence for a unit structure of 100 Å consisting of two 35 to 40 Å fibrils (the Watson-Crick *double helix* is approximately 20 Å wide). Intricate folding or coiling at size levels varying from 10s of Å to a few microns is present in metaphase chromosomes and various models for the ultrastructure have been proposed.[7]

9

MEIOTIC CHROMOSOMES (Fig. 1-7)

For several reasons human chromosomes during meiosis are less well characterized than during mitosis. Obtaining material for study usually presents a problem. Meiosis in the male can be studied in testicular biopsies after hypotonic treatment, fixation of the material, and staining with a DNA stain. In optimal material all stages of meiosis can be examined in one specimen, in particular pachytene (Fig. 1-8),

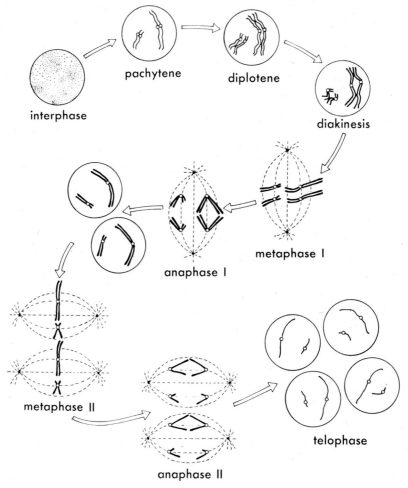

Figure 1-7. *Meiosis. Two pairs of chromosomes are shown. The original cell gives rise to 4 haploid cells (gametes). (From* An Introduction to Human Genetics *by H. Eldon Sutton. Copyright 1965 by Holt, Rinehart and Winston, Inc. Reproduced by permission of Holt, Rinehart and Winston, Inc.)*

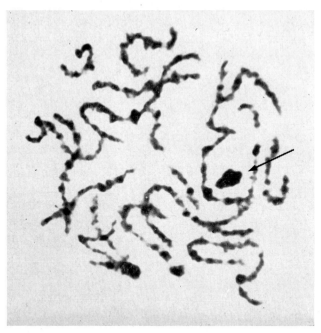

Figure 1-8. *Pachytene nucleus of a* spermatocyte *obtained by testicular biopsy of a 63-year-old man. The heavily condensed XY-bivalent is seen right center (arrow). Feulgen stain. (Courtesy of Dr. Susumu Ohno, Department of Biology, City of Hope Medical Center, Duarte, California, and reproduced with permission from* Human Chromosome Methodology, *Jorge J. Yunis, ed. Copyright 1965 by Academic Press Inc.)*

diakinesis, and metaphase I (Fig. 1-9).[13] At meiotic *metaphase I* the chromosomes are more condensed than at mitotic metaphase. The XY *bivalent* (pair) can be identified among the other 22 bivalents. In cases of infertility and in individuals with *nondiploid* chromosome complements, abnormal pairing as well as morphologically abnormal chromosomes are described. Meiotic chromosome examination is important for the understanding of such human reproductive problems as infertility and the *segregation* or passing on to the next generation of chromosome abnormalities.

Meiosis in the female is even more difficult to study (Fig. 1-10). In humans, the developmental sequence of *oögenesis*[11] is considerably different from that of *spermatogenesis*. *Oögonia* cease to propagate while the female is in the fifth month of fetal life, and all *oöcytes* complete the entire process of first meiotic prophase several weeks before

11

the end of gestation. Oöcytes then enter the long, interphase-like (dictyotene) stage during which the cytoplasm stores nutrients. At its shortest, the dictyotene stage lasts about 12 years; at its longest, about 50 years. When the menstrual cycle begins, one or more oöcytes at a time resume *meiosis I* from diakinesis shortly before ovulation. Thus, a meiotic metaphase I figure of an oöcyte can be obtained only from a mature follicle in the ovary of a woman in her reproductive period, on the day of ovulation. Meiosis II is usually completed only after an ovum has been fertilized. However, an ovulated but unfertilized ovum that has moved into the Fallopian tube will yield meiotic *metaphase II* figures.

Several points about human oögenesis deserve additional comment. During the long arrested stage in the middle of meiosis (dictyotene) the DNA is interphase-like in appearance and has not been studied ex-

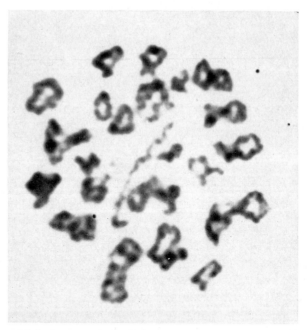

Figure 1-9. *Meiotic* metaphase I *figure of a* spermatocyte. *In addition to the 22* autosomal bivalents, *the X and Y seen in the center are in characteristic end-to-end association. Feulgen stain. (Courtesy of Dr. Susumu Ohno, Department of Biology, City of Hope Medical Center, Duarte, California, and reproduced with permission from* Human Chromosome Methodology, *Jorge J. Yunis, ed. Copyright 1965 by Academic Press Inc.)*

12

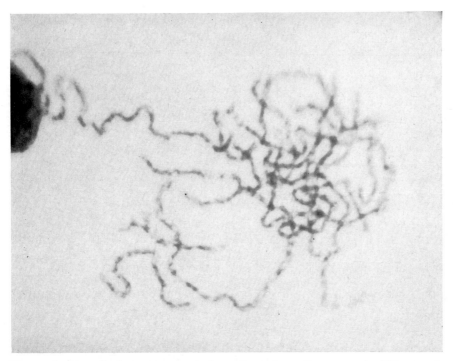

Figure 1-10. *Pachytene nucleus of an* oöcyte *from the ovary of a 5½-month-old female fetus. All 23* bivalents *demonstrate a fine* chromomeric *pattern. Feulgen stain. (Courtesy of Dr. Susumu Ohno, Department of Biology, City of Hope Medical Center, Duarte, California, and reproduced with permission from* Human Chromosome Methodology, *Jorge J. Yunis, ed. Copyright 1965 by Academic Press Inc.)*

tensively. Questions concerning what metabolic activities are going on and how susceptible the genetic material is to changes induced by x-ray and other environmental factors presently remain unanswered.

Because of the limitations inherent in *in vivo* studies of both male and female meiosis, *in vitro* techniques become all the more important. Mammalian testes can be grown in *organ culture.* In one study, testes of 14-day-old rats were grown for periods of up to six months.[14] Another interesting study reports resumption of post-dictyotene stages of meiosis in 80% of human oöcytes released from their follicles into a culture medium.[3]

Experimental control of chromosome segregation in meiosis is now reported, though not with mammalian chromosomes.[10] Direct mechanical micromanipulation of living meiotic chromosomes permits analysis of the mechanisms controlling *centromere* orientation and makes possible experimentally determined segregation of individual chromosomes.

INTERPHASE CHROMOSOMES AND SEX CHROMATIN

It is during interphase that chromosomes are the most interesting from a metabolic and genetic point of view. In this stage of many types of human diploid cells, DNA stains reveal small, scattered, darkly staining areas throughout the nucleus, which are referred to as *heterochromatin*. (A structure containing *DNA* is loosely referred to as *chromatin*.)

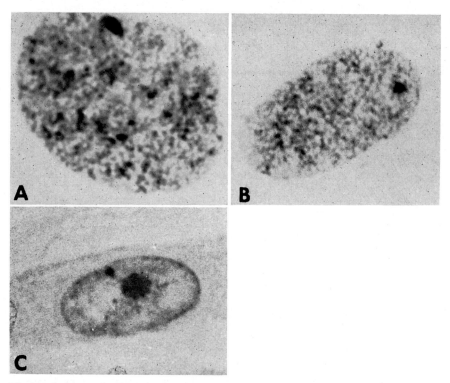

Figure 1-11. *A.* Sex chromatin *in tissue culture of a* diploid *human female. Position is near the nuclear membrane. Thionin stain. B. Human* sex chromatin *in tissue culture. Position is away from the nuclear membrane. Thionin stain. C. Human* sex chromatin *in a neuron. Thionin stain. In this nucleus the* sex chromatin *body is roughly triangular in shape. A single, large* nucleolus *is also present but does not stain as darkly as the* sex chromatin *when a* DNA *stain is used.*

A discrete, larger area of heterochromatin, often peripheral in location, is found in XX cells and is termed *sex chromatin* (Figs. 1-11 A,B,C). This heterochromatin represents a sizable portion of one X chromosome, while the less discrete and smaller DNA staining areas are probably *autosomal*. Although the genetic significance of heterochromatin is by no means fully clarified, connections with gene action have long been recognized. In mice (and other mammals, including humans) the *heterochromatization* of the X chromosome is associated with temporary inactivation of its genes.[9] A number of workers have shown in a variety of organisms that the DNA in *heterochromatic* regions of the nucleus replicates later than it does in *euchromatic* regions. In the nuclei of calf lymphocytes, up to 80% of the DNA is visible as heterochromatin and is in the "repressed" condition, so that it does not participate in the synthesis of *messenger RNA*. Furthermore, the repression of gene action that is characteristic of heterochromatin appears to be associated with the binding of DNA with basic protein or histone into a supramolecular complex.

It appears, then, that interphase nuclei of man contain both "active" and "repressed" DNA. The latter stains with DNA stains, as do metaphase chromosomes. Greater condensation of the DNA molecular complex is thought to account for the greater apparent affinity for DNA stain. Thus both interphase heterochromatin and metaphase chromosomes may be thought of as condensed DNA.

REFERENCES

1. Bajer, A.: Subchromatid Structure of Chromosomes in the Living State. Chromosoma, *17:*291, 1965.

2. Chicago Conference: Standardization in Human Cytogenetics. Birth Defects: Original Article Series, *II:*2, New York, The National Foundation, 1966.

3. Edwards, R. G.: Maturation *in vitro* of Human Ovarian Oöcytes. Lancet, *ii:*926, 1965.

4. Hayflick, L., and Moorhead, P. S.: The Serial Cultivation of Human Diploid Cell Strains. Exp. Cell Res., *25:*585, 1961.

5. Hayflick, L.: The Limited *in vitro* Lifetime of Human Diploid Cell Strains. Exp. Cell Res., *37:*614, 1965.

6. Hsu, T. C.: Mammalian Chromosomes *in vitro*. The Karyotype of Man. J. Hered., *43:*167, 1952.

7. Hyde, B. B.: Ultrastructure in Chromatin. Progr. Biophys., *15:* 129, 1965.

15

8. Ledley, R. S., and Ruddle, F. H.: Chromosome Analysis by Computer. Sci. Amer., *214*:40, 1966.

9. Lyon, M. F.: Sex Chromatin and Gene Action in the Mammalian X-Chromosome. Amer. J. Hum. Genet., *14*:135, 1962.

10. Niklas, R. B.: Experimental Control of Chromosome Segregation in Meiosis. J. Cell Biol., *27*:117A, 1965.

11. Ohno, S., Klinger, H. P., and Atkin, N. B.: Human Oögenesis. Cytogenetics, *1*:42, 1962.

12. Robins, J. H.: Tissue Culture Studies of the Human Lymphocyte. Science, *146*:1648, 1964.

13. Sasaki, M., and Makino, S.: The Meiotic Chromosomes of Man. Chromosoma, *16*:637, 1965.

14. Steinberger, A., Steinberger, E., and Perloff, W. H.: Mammalian Testes in Organ Culture. Exp. Cell Res., *36*:19, 1964.

15. Tjio, J. H., and Levan, A.: The Chromosome Number of Man. Hereditas, *42*:1, 1956.

2

Human Chromosome Variations and Abnormalities

After the correct description of the normal human chromosome number (1956) came the recognition of chromosome abnormalities clearly associated with specific congenital defects clinically. There came also a period of uncertainty, still not ended, concerning the significance of some chromosome abnormalities that are less clearly associated with specific congenital defects and may, in fact, be unrelated to any clinical abnormality. Table 2-1 presents a summary of human chromosome variations and abnormalities clearly associated with or clearly not regularly associated with clinical abnormality. In the table the classification of clinically significant chromosome abnormalities is based on whether

Table 2-1. Human Chromosome Variations and Abnormalities

A. Variations not regularly associated with any clinical abnormality.
  1. Some variations in chromosome length
  2. Enlarged satellites
  3. Variations in expression of secondary constrictions
  4. Some cases of mosaicism
  5. "Balanced" structural rearrangements such as translocation chromosome carriers

18

B. Abnormalities of sex chromosomes associated with clinical abnormality. (The number of sex chromatin bodies in each case is one less than the number of X chromosomes.)
  1. Klinefelter syndrome
  2. Turner syndrome (gonadal dysgenesis)
  3. Multiple X females
  4. Intersexes
C. Abnormalities of the autosomes associated with clinical abnormality.
  1. D group trisomy or 13-15 trisomy
  2. E group trisomy or 18 trisomy
  3. G group trisomy, 21 trisomy, Down's syndrome, or mongolism (including both standard trisomy mongolism and translocation mongolism)
  4. B chromosome deletion
  5. 18 deletion
D. Abnormalities associated with neoplasms.

sex chromosomes or *autosomes* are involved. In some cases both may be involved, as in *triploidy* (three instead of two chromosomes in each pair) and in multiple types of *aneuploidy (nondiploidy)* occurring in the same individual, such as in Klinefelter-mongol. Other classifications are

19

possible. An example is a division according to whether number or morphology is primarily affected. Since these two may overlap considerably, an alternative is to merely list the type of abnormality, such as *trisomy, monosomy, polyploid, triploid, aneuploid, mosaic, fragment, translocation, deletion, ring chromosome, isochromosome.*

HUMAN CHROMOSOME VARIATIONS (Table 2-1A)

1. LENGTH. Some of the variation in chromosome *length* may be accounted for by artifact introduced during preparation of chromosomes for examination. *Homologous* chromosomes, for instance, are rarely the same length. *Chromatids* of the same chromosome *arm* are rarely the same length. Photographs of chromosomes from which arm measurements are made can be out of focus. Other examples of variations in chromosome length are thought to reflect true *in vivo* differences more than preparation artifact, though the distinction is often difficult. Between 2 and 3% of normal adults have structural anomalies confined to one autosome.[7] Among males, between 2 and 3% of subjects show well-marked variations in the length of the Y chromosome.[26] The X chromosome in female cells, which at interphase composes the sex chromatin, at metaphase has been found to vary considerably in total length, *arm ratio,* and *centromere index.*[2, 35]

Where this normal variation stops and clinically significant decrease or increase of chromosome material starts, is sometimes difficult to tell. Several types of large deletions, associated with clinical abnormality, are well described. One is the B chromosome deletion syndrome to be discussed again later. A very small, presumably *deleted* G group chromosome called the Philadelphia chromosome is present in chronic myelogenous leukemia. Deleted X chromosomes are described in individuals who resemble clinically individuals missing an entire X chromosome. Much larger than normal chromosomes associated with clinical abnormality are also well described. An isochromosome for the long arm of the X chromosome is grossly above the normal size range and is found in abnormal individuals. Translocation chromosomes may be grossly larger.

2. SATELLITES. *Satellites* can be present normally on all of the D and G group metaphase chromosomes.[12] Even the most excellent preparations rarely reveal all of these 10 chromosomes to be *satellited* in the same cell. Techniques of preparation are usually thought to account for most of the variable presence of satellites. Satellited chromosomes

20

are often seen to be *associated* in chromosome preparations from normal individuals. *Association* in this context means that the chromosomes tend to group together with their satellites close together. Enlarged satellites were at first thought to be related to clinical abnormality but now are known to be present in normal individuals and may be inherited.[6]

3. SECONDARY CONSTRICTIONS.   The significance of *secondary constrictions* or nonstaining areas along the chromatid arms is not entirely clear. It is well known that changes in appearance of secondary constrictions can be induced by changes in techniques of preparation of the chromosomes. However, secondary constrictions are consistently present in certain portions of particular chromosomes and are utilized in the standard classification of normal chromosomes. Areas consistently involved are the proximal region of the long arm of chromosome No. 1, the proximal part of the long arm of No. 16, and the long arm of the Y chromosome.   As yet there is not good evidence for a connection between

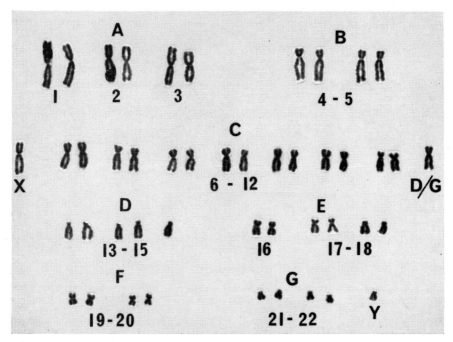

Figure 2-1.   *The* karyotype *found in D/G translocation mongolism. 46 chromosomes are present with 1 normal chromosome 15 and 2 normal chromosomes 21, but in addition there is a large chromosome formed by* translocation *between chromosomes 15 and 21.*

variation of secondary constrictions and gross *phenotypic* variations in humans. Some agents known to induce chromosome aberrations cause *breaks* preferentially in regions of secondary constrictions.

4. MOSAICISM. Genetic *mosaicism* or the mixture in one organism of two or more genetically unlike cell lines is well known in many forms of life and for various reasons has survived through evolution. All humans are probably chromosomal mosaics in the sense that, in addition to *diploid* cells, some cells are *nondiploid*. If the abnormal cells are of sufficient number or if they involve critical tissues such as the gonads, we become aware of the phenotypic effects. Blood is the tissue usually sampled for chromosome study for reasons already explained, and a complete evaluation for mosaicism in human beings is impossible and unnecessary. Nevertheless, efforts to evaluate a limited number of different tissues become important in certain clinical situations when the diagnosis is critical or predictions need to be made for the offspring.

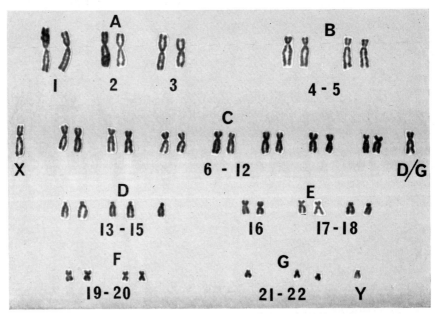

Figure 2-2. *The* karyotype *found in the carrier of translocation mongolism. The carrier is* phenotypically *normal but has 45 chromosomes, one of which is a fused chromosome. One normal chromosome 15 and 1 normal chromosome 21 are present in this karyotype. The* translocation *represented here is a D/G translocation. G/G translocations may also be associated with mongolism.*

5. BALANCED STRUCTURAL REARRANGEMENT. Perhaps the most studied example of a *"balanced" structural rearrangement* is the carrier of translocation mongolism. This individual is typically a young mother (phenotypically normal) who delivers a mongol child (see Table 2-4). There may also be a history of mongolism in her family. When chromosome examinations are made, the mongol child is found to have 46 chromosomes, but one of these is abnormal in appearance and is a translocation chromosome (Fig. 2-1). A clinically normal parent is found to have only 45 chromosomes, including the translocation chromosome (Fig. 2-2). The most common type of translocation involves joining of a D and a G chromosome (D/G) and is discussed here. Important principles to remember are: (1) The translocation is "balanced" in the carrier and does not represent enough change in functional genetic material to alter the *phenotype*. (2) In the mongol child, three instead of two G group chromosome equivalents are present because the translocation chromosome carries the genetic equivalent of a G group chromosome and in addition the two other chromosomes in this G group pair are present.

HUMAN CHROMOSOME ABNORMALITIES

Sex Chromosomes

Clinically significant chromosome abnormalities involving the sex chromosomes will now be discussed (Table 2-1, B; also see Chapter 3). Terminology in this area is less than satisfactory since there is considerable carry-over of syndrome names used prior to discovery of the chromosome abnormality.

1. KLINEFELTER SYNDROME. The Klinefelter syndrome is used to name the group of *phenotypic* males, usually sterile, with small testes, who have varying clinical stigmata of the syndrome described by Klinefelter, Reifenstein, and Albright.[21] Some of these patients were found to be *chromatin positive*.[11] A case of *sex chromatin positive* Klinefelter syndrome was subsequently shown to have an XXY chromosomal complement[20] (Fig. 2-3). Following demonstration of an extra X chromosome in addition to a Y in somatic cells of the "prototype" Klinefelter, a wide spectrum of sex chromosome abnormalities was subsequently added to the list (see Table 2-2, A for a partial listing). The correlation of the phenotypic spectrum of this group with the *genotypic* spectrum is not complete.

23

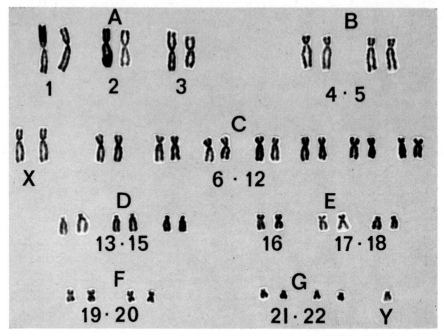

Figure 2-3. *The* karyotype *found in Klinefelter syndrome. 47 chromosomes are present, including 2 X chromosomes and a Y chromosome.*

Table 2-2. Variants of Sex Chromosome Abnormalities (The number of sex chromatin bodies in each case is one less than the number of X chromosomes.)

Sex chromosome complement	Total chromosome number
A. *Klinefelter syndrome*	
XXY	47
XXXY	48
XXXXY	49
XXYY	48
XXXYY	49
XY	46
XYY	47
XXY/XX	47/46
XXY/XY	47/46
XXY/XXXY	47/48
XXXY/XXXXY	48/49

B. *Turner syndrome* (gonadal dysgenesis)

XO	45
XO/XX	45/46
XO/XX/XXX	45/46/47
XO/XXX	45/47
XX^I (I = isochromosome)	46
XO/XX^I	45/46
XX^D (D = deleted)	46
XO/XX^D	45/46
XS (S = small abnormal sex chromosome)	46
XO/XS	45/46
XO/XY	45/46
XX	46
XY	46

C. *Multiple X females* (non-Turner phenotypes)

XXX	47
XXXX	48
XXXXX	49
XX/XXX	46/47
XO/XXXX	45/48

D. *Intersexes*

XX/XY	46
XX	46
XY	46
XO/XY	45/46
XO/XYY	45/47
XO/XX/XY	45/46
XX/XXY/XXYYY	46/47/49
XO/XX^{D}Y	45/47

A subtype deserves special comment, the XYY or double Y chromosome complement. These phenotypic males are usually tall and aggressive and may be either mentally retarded or mentally ill and criminally insane. Chromosome studies on 129 tall men surveyed in four different institutions for the care of criminal males in Pennsylvania (U.S.A.) showed that 1 in 11 subjects displayed *aneuploidy* of the sex chromosomes. (5 were 47, XYY and 7 were sex chromatin positive.)[41]

2. TURNER SYNDROME. The Turner syndrome (or gonadal dysgenesis) includes those individuals with a *phenotypic spectrum* from female to male, who have varying clinical stigmata of the syndrome described by Turner[42] (Table 2-3). It was shown that some patients with Turner syndrome are *chromatin negative*.[33] An XO chromosome constitution was subsequently demonstrated in a typical patient with female

Table 2-3. Characteristics of Turner Syndrome (Gonadal Dysgenesis)

Short stature
Shield chest
Webbed neck
Lymphedema
Short metacarpal IV
Hypoplastic nails
Pigmented nevi
Congenital heart disease
Failure of menstruation
Streak gonads
Masculinization (may be present)
Mental defect

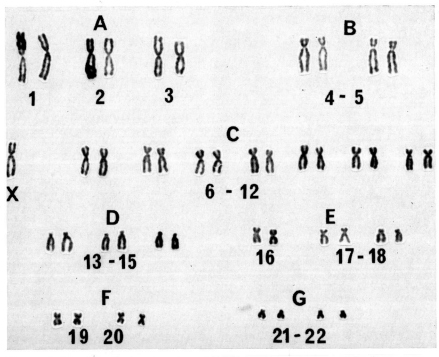

Figure 2-4. *The* karyotype *found in Turner syndrome. 45 chromosomes are present, including only 1 instead of 2 X chromosomes.*

phenotype[14] (Fig. 2-4). Following demonstration of a missing X chromosome in *somatic* cells of the "prototype" Turner, a wide spectrum of sex chromosome abnormalities was subsequently added to the list (see Table 2-2, B for a partial listing).

3. MULTIPLE X FEMALES. Classified separately from the Turner group are apparent females who do not have the clinical Turner syndrome just described. There may be some signs of masculinization and occasional clinical overlap with the Turner group. An XXX chromosome constitution was first demonstrated to be the prototype of this group[19] (Fig. 2-5). In the same report two *sex chromatin* bodies were found in some cells from buccal smears (Fig. 2-6). Subsequently a spectrum of sex chromosome abnormalities has been added (Table 2-2, C). For the most part, the chromosome abnormalities represent additions of X chromosome complements, beyond XX. Phenotypic and genotypic correlations are the most difficult to make in this group. The original XXX case had failed to menstruate, and ovarian biopsy revealed

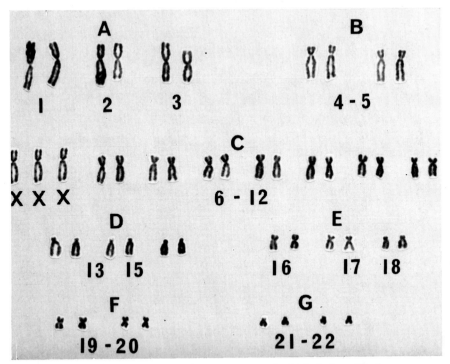

Figure 2-5. *The karyotype found in the triple X female. 47 chromosomes are present, including 3 X chromosomes.*

27

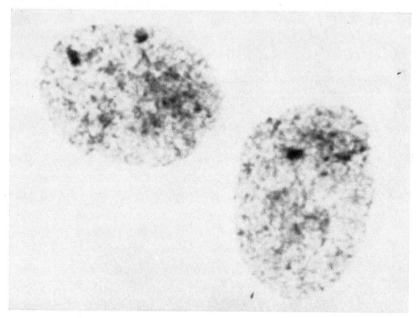

Figure 2-6. *An illustration of double* sex chromatin *bodies.*

deficient follicle formation. Other individuals confirmed to have extra X chromosomes have been fertile. Interestingly enough, none of the offspring has had extra chromosomes, suggesting selection against *ova* or early *zygotes* containing additional X chromosome material.

The incidence of *chromatin negative* females at birth is approximately 0.4 per 1000 females, according to one study.[25] This incidence is considerably lower than the incidence of 1.96 chromatin positive males per 1000 males or 1.2 double chromatin positive females per 1000 females in the same study. It is suggested that XO individuals may be less viable than XXY or XXX individuals.

4. INTERSEXES. Another group of clinically significant sex chromosome abnormalities remains to be mentioned. This group includes *hermaphrodites* who show varying development of both male and female genital systems (Table 2-2, D). It should be said here that many intersex individuals show no evidence of sex chromosome abnormalities. If a prototype of the intersex group with abnormal sex chromosomes is to be picked, it should be the XX/XY mosaic genotype with both male and female chromosomes.

28

Autosomes

Clinically significant chromosome abnormalities involving the autosomes will now be considered. There are four well-established syndromes with consistent phenotypes (Table 2-1, C). These are D trisomy[31, 40] (Fig. 2-7), E (18) trisomy[10] (Fig. 2-8), G trisomy or mongolism[22, 24, 32] (Fig. 2-9), and B chromosome deletion[15, 23] (Fig. 2-10).
Deletions of chromosome 18 also appear to be associated with a fairly
characteristic clinical picture.[9] *Autoradiography* following label with
tritiated thymidine has provided a means of classifying the chromosome
pairs involved in D and G trisomy and B chromosome deletion syndrome, since the pairs in these groups cannot be distinguished morphologically (see Chapter 3 and Table 3-7). From Tables 2-4 and 2-5
the phenotypes in these syndromes can be compared. It is interesting
to note both similarities and differences. What parts of the syndromes
will turn out to be directly related to genes on the involved chromosomes,

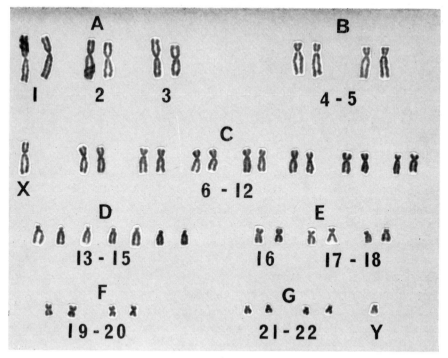

Figure 2-7. *The* karyotype *seen in D trisomy. 47 chromosomes are present, including 7 instead of 6 chromosomes in the D group.*

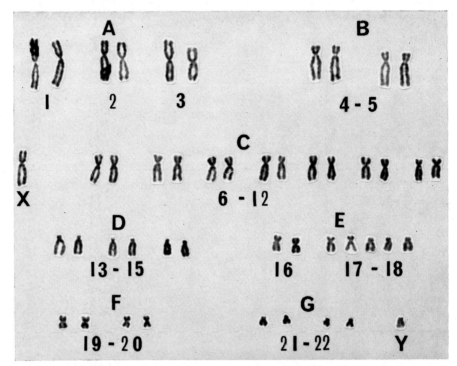

Figure 2-8. *The* karyotype *seen in E trisomy. 47 chromosomes are present, including 7 instead of 6 chromosomes in the E group. The extra chromosome is represented as belonging to pair 18.*

what parts are related to a more general "imbalance" of genetic material, and what parts are etiologically unrelated to the chromosome abnormality are at present unclear. Efforts, for instance, to demonstrate quantitatively the effects of three homologous genes instead of two have met with poor success in G trisomy. One problem is to know which gene products to measure, since the autosomes are, for all practical purposes, not mapped.

Chromosome variants of the autosomal trisomy syndromes exist. These include mosaicism, translocation, *duplication* chromosomes, and ring chromosomes. Translocation G trisomy (mongolism) has already been mentioned under the discussion of translocation carriers.

A comparison of sex chromosome and autosome abnormalities suggests that much less variation of autosomes (in terms of chromosome material visible at metaphase) is tolerated in humans. However, both sex chromosome and autosome abnormalities are found in spontaneous abortions. In one large study, among 200 specimens, 44 were found to have chromosome abnormalities, including XO, XXY, XYY, XXX, XXXX, triploid XXX, *tetraploid* XXXX, *tetraploid* XXYY, pair 3 trisomy, pair 16 trisomy, and B, C, D, E (18), and G group trisomies.[3]

Neoplasms

Direct chromosome analysis of a bone marrow sample is useful for the diagnosis of chronic myelogenous leukemia when a deleted G group chromosome is found (Philadelphia chromosome, Ph[1] chromosome)[29] (Fig. 2-11) and may precede definitive clinical diagnosis by other diagnostic tests. No other specific chromosome abnormalities are found to be associated consistently with leukemias. However, marrow chromosome changes of more than one type have been found to be useful in the diagnosis of preleukemia.[30] Chromosome abnormalities of different types may be present in direct tissue specimens, in fluid effusions, or in tissue cultures from various types of human solid tumors.

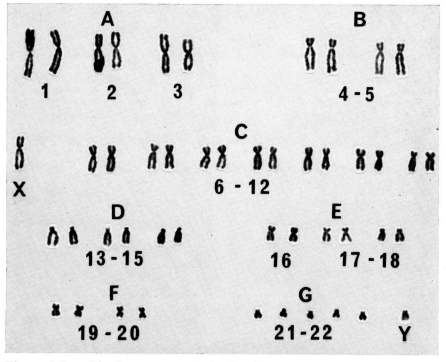

Figure 2-9. *The* karyotype *seen in G trisomy (mongolism). 47 chromosomes are present, including 5 instead of 4 chromosomes in the G group. This is a male* karyotype *and the Y chromosome resembles the G group.*

31

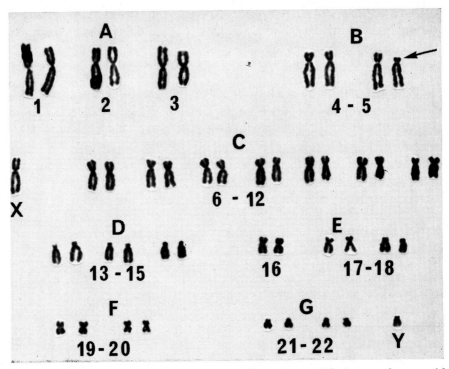

Figure 2-10. *The* karyotype *seen in B chromosome deletion syndrome. 46 chromosomes are present, but 1 chromosome in the B group (arrow), represented here as a pair 5 chromosome, has a deletion of the short arms.*

Table 2-4. Phenotypes of the Autosomal Trisomy Syndromes

	D Group Trisomy 13-15 Trisomy	E Group Trisomy 18 Trisomy	G Group Trisomy 21 Trisomy Down's Syndrome Mongolism
Growth	Failure to thrive	Failure to thrive Av. birth wt. less than 5½ lbs. Gross retardation of skeletal maturation	Small stature
Nervous System	Mental retardation Apparent deafness Seizures	Mental retardation Hypertonicity (stiff)	Mental retardation Hypotonicity (limp)

32

Table 2-4. *(Cont.)*

	D Group Trisomy 13-15 Trisomy	E Group Trisomy 18 Trisomy	G Group Trisomy 21 Trisomy Down's Syndrome Mongolism
Head	Sloping forehead	Prominent occiput	Flat occiput Small
Eyes	Small or abnormal		Slanting Speckling of iris
Ears	Low set, malformed	Low set, malformed	Low set, malformed
Mouth	Cleft palate Cleft lip	Small chin	Dental malformations Large, furrowed tongue
Hands and Feet	Posterior prominence of heels Flexion of fingers Extra digits	Flexion of fingers "Rocker bottom" feet	Short, broad hands Short fingers Curved 5th finger Gap between toes 1 and 2
Skeletal		Short sternum Small pelvis	Hyperextensible joints Broad, short neck
Cardiac	Congenital heart defects	Congenital heart defects	Congenital heart defects
Abdominal	Hernias Defective kidneys	Hernias Defective kidneys	Hernias
Genitalia	Abnormal	Abnormal	Underdeveloped
Prognosis	Poor	Poor	Fair—life expectancy shorter than average
Directly Related to Maternal Age	Yes	Yes	Yes
Sex	60% female	70% female	No sex difference
Incidence at Birth	0.45/1000	0.23/1000	1.6/1000
Other Congenital Malformations	Frequent	Frequent	Frequent

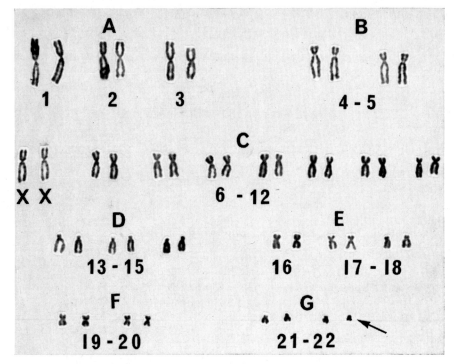

Figure 2-11. *The* karyotype *seen in chronic myelogenous leukemia. A G group chromosome with* deleted *long* arms *(arrow) is consistently seen and is referred to as the Philadelphia (Ph[1]) chromosome.*

Table 2-5. Characteristics of the B Chromosome Deletion Syndrome (Cri du Chat syndrome)

Characteristic cry (like a cat)
Mental retardation
Growth retardation
Small head
Rounded faces
Wide eyes
Low set ears
Small chin
Crossed eyes

34

ETIOLOGY OF CHROMOSOMAL ABNORMALITIES[8]

Maternal age

The increasing risk for mongol offspring (G trisomy) after maternal age 35 has long been known. Increased maternal age is also reported for D and E trisomy and for Klinefelter syndrome, although the number of cases is by no means as great as for G trisomy. Because no clear association between chromosome abnormalities and paternal age is reported, etiologic factors pertaining to the female germ cells or the uterine environment of the early zygote are suggested.

Translocation carriers as a cause of mongolism have already been discussed under chromosome variations. Translocation mongolism is more likely to be found by examining the chromosomes of mongols born of young mothers, than by examining a random group of mongols. The reason is that increasing maternal age is by far the most frequent single factor associated with the production of offspring with clinical mongolism. Young mothers do not reflect this etiologic factor and therefore are more likely to reflect other factors, in particular, the translocation carrier state.

Irradiation

Breaks (Fig. 2-12) and abnormal chromosome recombinations have been demonstrated following exposure of single human cells in culture

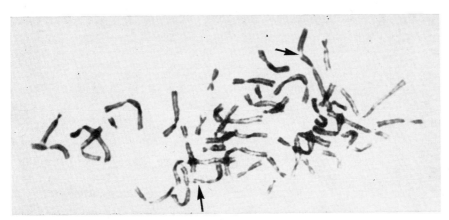

Figure 2-12. *The upper arrow illustrates a* chromatid break. *The staining of the* chromatid arm *is interrupted, and in addition, the* chromatid *alignment is disturbed. The lower arrow illustrates a* chromatid gap. *The staining is interrupted but the alignment is not.*

35

to x rays.[36] The average dose needed to produce one break per cell is as low as 20 to 25 roentgens if the breaks are scored immediately after irradiation. Puck states that mammalian *somatic* cells *in vitro* are exceedingly sensitive to reproductive killing by x-irradiation. He reviews evidence that the cell lesion responsible for reproductive death is lodged in the chromosomes.[37]

There are many reports of chromosome aberrations in human beings following diagnostic and therapeutic x ray and accidental exposure to ionizing irradiation. Persistence of these aberrations in peripheral blood cells for 3½ years is reported.[1] The incidence of certain human tumors is related to exposure to irradiation, including thyroid carcinoma, bone tumors, skin cancers, and some kinds of leukemia. One study of irradiation exposure in parents of children with mongolism revealed a statistically significant association between maternal ionizing radiation exposure and mongolism. In contrast, there were no significant differences in ionizing radiation exposure in the fathers of the mongol and control children. However, a surprising increase in radar exposure was discovered in a significant number of fathers of the mongol cases.[39]

Viruses

Viruses can produce many types of change involving *in vitro* chromosomes, including *breaks* (Fig. 2-12), *pulverization* (Fig. 2-13), *exchange figures* (Fig. 2-14), cell fusion, and *transformation*.[28] How specific the gross chromosome change is to the inducing agent is unclear. For instance, irradiation (and other agents as well) can produce breaks morphologically indistinguishable by light microscopy from breaks induced by viruses. On the other hand, Moorhead and Saksela have suggested that the preferential effect of certain viruses on certain regions of the chromosome could imply that sequences of *DNA* in these regions are similar to or in some way related to those of the virus.[27] These same authors also suggest the possibility of incorporation of the virus itself into susceptible chromosome regions where it could, say, act as promoter for fusion with other *chromatin* segments.

The role of viruses in producing cell fusion and transformation in tissue culture should be mentioned. Nondiploid cells from mouse and man have been fused, by the introduction of Sendai virus, to form *heterokaryons*.[16] Human diploid cells undergo transformation (with *diploid* to *nondiploid* chromosome change) when infected with Simian virus 40 (SV40), and at the same time the cells change from a population

36

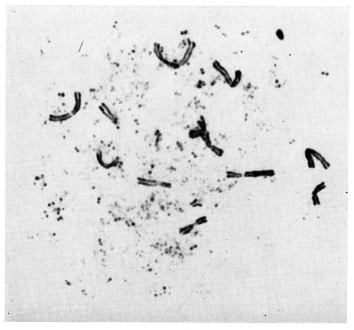

Figure 2-13. *An example of chromosome* pulverization. *Some chromosomes can be identified, but the remainder are represented by small* DNA *staining masses of various sizes.*

with limited life expectancy to one with unlimited life expectancy in culture.

To summarize partially *in vivo* and epidemiological evidence of virus induced chromosome change, the different abnormalities of chromosomes already described have been reported, usually in cultured leukocytes, in patients with infectious mononucleosis, mumps, German measles, chicken pox, and other less specifically diagnosed but presumed viral exanthems and meningitis. Grossly damaged chromosomes were also found in cultured leukocytes from subjects vaccinated against yellow fever, with live attenuated virus. In a newborn sex chromatin screening study, "an incidence of 0.6% of sex chromatin aberrations in newborns was found during a 5-month period, while no aberrations occurred in similar populations before and after this period. Down's syndrome also exhibited an elevated frequency during this same critical interval. A severe rubella epidemic may have influenced this pattern."[38]

37

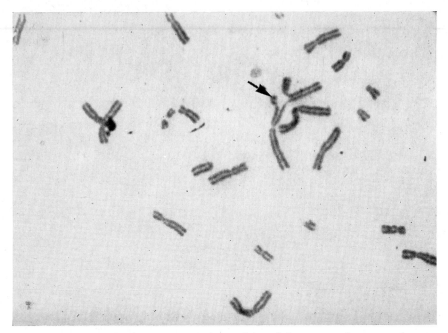

Figure 2-14. *An* exchange *figure (arrow) involving more than one chromosome. This type of chromosome association at metaphase is considered abnormal in somatic cells.*

Drugs

Various kinds of drugs, chemicals, and antimetabolites can affect chromosomes, producing some or all of the defects already mentioned. To more detailed examination interesting differences as well as similarities between the effects of various agents on chromosomes suggest ways to investigate principles of chromosome structure and function. A few examples will be discussed here. Mammalian chromosomes in culture treated with bromodeoxyuridine (BUdR, an analogue which is incorporated into *DNA* in place of thymidine) show marked lengthening of centromeric regions and *secondary constriction* areas, as well as an increase in the frequency of *chromatid breakage,* especially at sites of secondary constrictions.[18] Mitomycin C, an antibiotic known to disrupt DNA metabolism, causes breaks and *exchanges* in the chromosomes of cultured human leukocytes.[5] There is a marked excess of breaks in the secondary constriction regions of chromosome Nos. 1, 9, and 16. Therefore, Mitomycin C and BUdR show preferential action on certain chromosome areas. Irradiation, on the other hand, does not show this preferential action. Why should these chromosome points be "weak" and what other agents act on the same or different "weak" points? What relation is there to detailed mechanisms of action of these agents in DNA metabolism?

Autoimmunity[13]

There is evidence indicating an association between thyroid *auto-immunity* and chromosomal abnormalities, in particular Down's syndrome and Turner syndrome. Unsettled is the question of whether some other more basic process may not underlie both the autoimmunity and the chromosome aberration. A possible association of mother-child incompatibilities for ABO and Rh blood group alleles, with certain types of chromosomal aberrations has been reported.[4]

Familial tendency

Familial clusters of chromosome abnormalities of the same or different types occur and suggest genetic tendencies but do not rule out or clarify the importance of environmental events.[17] It is possible that an environmental event such as irradiation could potentiate a genetic tendency. Another familial mechanism, at the chromosome rather than the gene level, is reproduction of chromosomally abnormal individuals. The occurrence of balanced translocation carriers in families and the increased risk for chromosomally abnormal offspring has already been discussed. Mongol females are known to produce both mongol and normal offspring.[34]

REFERENCES

1. Bender, M. A., and Gooch, P. C.: Persistent Chromosome Aberrations in Irradiated Human Subjects. II. Three and One-Half Year Investigation. Radiat. Res., *18:*389, 1963.
2. Bishop, A., Leese, M., and Blank, C. E.: The Relative Length and Arm Ratio of the Human Late-Replicating X Chromosome. J. Med. Genet., *2:*107, 1965.
3. Carr, D. H.: Chromosome Studies in Spontaneous Abortions. Obstet. Gynec. (NY), *26:*308, 1965.
4. Chandra, H. S.: Mother-Child Incompatibilities for ABO and Rh Alleles. Possible Association with Certain Types of Chromosomal Aberrations. New Eng. J. Med., *272:*566, 1965.
5. Cohen, M. M., and Shaw, M. W.: Effects of Mitomycin C on Human Chromosomes. J. Cell Biol., *23:*386, 1964.
6. Cooper, H. L., and Hirschhorn, K.: Enlarged Satellites as a Familial Chromosome Marker. Amer. J. Hum. Genet., *14:*107, 1962.

7. Court Brown, W. M., Jacobs, P. A., Buckton, E. E., Tough, I. M., Kuenssberg, E. V., and Knox, J. D. E.: *Chromosome Studies on Adults.* London, Cambridge University Press, 1966.

8. Day, R. W.: The Epidemiology of Chromosome Aberrations. Amer. J. Hum. Genet., *18:*70, 1966.

9. de Grouchy, J.: Chromosome 18: A Topologic Approach. J. Pediat., *66:*414, 1965.

10. Edwards, J. H., Harnden, D. G., Cameron, A. H., Crosse, V. M., and Wolff, O. H.: A New Trisomic Syndrome. Lancet, *i:*787, 1960.

11. Ferguson-Smith, M. A.: Klinefelter's Syndrome: Frequency and Testicular Morphology in Relation to Nuclear Sex. Lancet, *ii:*167, 1957.

12. Ferguson-Smith, M. A., and Handmaker, S. D.: Observations on the Satellited Human Chromosomes. Lancet, *i:*638, 1961.

13. Fialkow, P. J.: Autoimmunity and Chromosomal Aberrations. Amer. J. Hum. Genet., *18*:93, 1966.

14. Ford, C. E., Jones, K. W., Polani, P. E., deAlmeida, J. C., and Briggs, J. H.: A Sex-Chromosome Anomaly in a Case of Gonadal Dysgenesis (Turner's Syndrome). Lancet, *i:*711, 1959.

15. German, J., Lejeune, J., Macintyre, M. N., and de Grouchy, J.: Chromosomal Autoradiography in the Cri du Chat Syndrome. Cytogenetics, *3*:347, 1964.

16. Harris, H., and Watkins, J. F.: Hybrid Cells Derived from Mouse and Man: Artificial Heterokaryons of Mammalian Cells from Different Species. Nature, *205*:640, 1965.

17. Hecht, F. H., Bryant, J. S., Gruber, D., and Townes, P. L.: The Nonrandomness of Chromosomal Abnormalities. Association of Trisomy 18 and Down's Syndrome. New Eng. J. Med., *271:*1081, 1964.

18. Hsu, T. C., and Somers, C. E.: Effect of 5-Bromodeoxyuridine on Mammalian Chromosomes. Proc. Nat. Acad. Sci., USA, *47:*396, 1961.

19. Jacobs, P. A., Baikie, A. G., Court Brown, W. M., MacGregor, T. N., Maclean, N., and Harnden, D. G.: Evidence for the Existence of the Human "Super Female." Lancet, *ii:*423, 1959.

20. Jacobs, P. A., and Strong, J. A.: A Case of Human Intersexuality Having a Possible XXY Sex-Determining Mechanism. Nature, *183:*302, 1959.

21. Klinefelter, H. F., Reifenstein, E. C., and Albright, F.: Syndrome Characterized by Gynecomastia, Aspermatogenesis without A-Lydigism, and Increased Excretion of Follicle-Stimulating Hormone. J. Clin. Endocr., *2:*615, 1942.

22. Lejeune, J., Gautier, M., and Turpin, R.: Étude des Chromosomes Somatiques de Neuf Enfants Mongoliens. C. R. Acad. Sci. (Paris), *248:*1721, 1959.

23. Lejeune, J., Lafourcad, J., Berger, R., Vialette, J., Boeswillwald, M., Seringe, P., and Turpin, R.: Trois Cas de Deletion Partielle des Bras Courts d'un Chromosome 5. C. R. Acad. Sci. (Paris), *257:*3098, 1963.

24. Lejeune, J.: The 21 Trisomy—Current Stage of Chromosomal Research. In: *Progress in Medical Genetics,* Vol. III, A. G. Steinberg, ed. New York, Grune and Stratton, 1964, p. 144.

25. Maclean, N., Harnden, D. G., Court Brown, W. M., Bond, J., and Mantle, D. J.: Sex Chromosome Abnormalities in Newborn Babies. Lancet, *i:*286, 1964.

26. Makino, S., Sasaki, M. S., Yamada, K., and Kajii, T.: A Long Y Chromosome in Man. Chromosoma, *14:*154, 1963.

27. Moorhead, P. S., and Saksela, E.: The Sequence of Chromosome Aberrations During SV40 Transformation of a Human Diploid Cell Strain. Hereditas, *52:*271, 1965.

28. Nichols, W. W.: The Role of Viruses in the Etiology of Chromosomal Abnormalities. Amer. J. Hum. Genet., *18:*81, 1966.

29. Nowell, P. C., and Hungerford, D. A.: Chromosome Studies in Human Leukemia. II. Chronic Granulocytic Leukemia. J. Nat. Cancer Inst., *27:* 1013, 1961.

30. Nowell, P. C.: Prognostic Value of Marrow Chromosome Studies in Human "Preleukemia." Arch. Path., *80:*205, 1965.

31. Patau, K., Smith, D., Therman, E., Inhorn, S. L., and Wagner, H. P.: Multiple Congenital Anomalies Caused by an Extra Autosome. Lancet, *i:*790, 1960.

32. Penrose, L. S., and Smith, G. F.: *Down's Anomaly*. Boston, Little, Brown and Company, 1966.

33. Polani, P. E., Hunter, W. F., and Lennox, B.: Chromosomal Sex in Turner's Syndrome with Coarctation of the Aorta. Lancet, *ii:*120, 1954.

34. Priest, J. H., Thuline, H. C., Norby, D. E., LaVeck, G. D.: Reproduction in Human Autosomal Trisomics. Amer. J. Dis. Child., *105:*31, 1963.

35. Priest, J. H., Heady, J. E., and Priest, R. E.: Delayed Onset of Replication of Human X Chromosomes. J. Cell Biol., *35:*483, 1967.

36. Puck, T. T.: Radiation and the Human Cell. Sci. Amer., *202:*142, 1960.

37. Puck, T. T.: Cellular Interpretation of Aspects of the Acute Mammalian Radiation Syndrome. In: *Cytogenetics of Cells in Culture,* R. C. J. Harris, ed. New York, Academic Press, 1964, p. 63.

38. Robinson, A., and Puck, T. T.: Sex Chromatin in Newborns: Presumptive Evidence for External Factors in Human Nondisjunction. Science, *148:*83, 1965.

39. Sigler, A. T., Lilienfeld, A. M., Cohen, B. H., and Westlake, J. E.: Radiation Exposure in Parents of Children with Mongolism (Down's Syndrome). Bull. Hopkins Hosp., *117:*374, 1965.

40. Smith, D. W., Patau, K., Therman, E., Inhorn, S. L., and DeMars, R. I.: The D_1 Trisomy Syndrome. J. Pediat., *63:*326, 1963.

41. Telfer, M. A., Baker, D., Clark, G. R., and Richardson, C. E.: Incidence of Gross Chromosomal Errors among Tall Criminal American Males. Science, *159:*1249, 1968.

42. Turner, H. H.: A Syndrome of Infantilism, Congenital Webbed Neck, and Cubitus Valgus. Endocrinology, *23:*566, 1938.

3

Nomenclature and Identification of Normal and Abnormal Human Mitotic Chromosomes

A number of tables are assembled in this chapter to summarize the nomenclature of human chromosomes. Agreements concerning nomenclature have been the result of three main conferences, in Denver (1960),[3] in London (1963),[6] and in Chicago (1966).[2] Texts that include further discussion of chromosome identification are listed at the end of this chapter.[1, 7, 8]

Table 3-1. Conspectus of Normal Human Mitotic Chromosomes. (Modified from the Denver Classification, 1960,[3] and the London Conference, 1963.[6])

General	The autosomes are serially numbered 1 to 22 as nearly as possible in descending order of length. The sex chromosomes are X and Y. The 22 autosomes are classified into 7 groups, A to G. The total number of chromosomes (2n, diploid) is 46 and is composed of 23 pairs, 22 of autosomes and 1 of sex chromosomes. The autosome pairs individually are morphologically homologous, as is the XX female sex chromosome pair. The XY male sex chromosome pair is morphologically nonhomologous. Prior to mitosis the chromatids replicate; consequently, the 46 chromosomes observed in mitoses arrested with colchicine are composed of pairs of chromatids joined at their centromeres. The centromere position is also used in the normal classification.

Group 1-3 (A)	These are the longest chromosomes and are readily distinguished, one from the other.

No. 1. Longest chromosome, median centromere. A secondary constriction may be observed in the proximal region of one arm (long).

No. 2. Next longest chromosome, submedian centromere.

No. 3. Third longest chromosome, median centromere.

Group 4-5 (B)	Long chromosomes with submedian centromeres. The two chromosome pairs are difficult to distinguish, but No. 4 is slightly longer.

Group 6-12 (C plus X)	Medium-sized chromosomes. Individual members of this group are difficult to distinguish. It may be best to assemble this group last when preparing karyotypes. In females there should be 16 members; in males, 15.

X, Nos. 6, 7, 8, 11 Comparatively median centromeres. One X in XX cells is late replicating.

Nos. 9, 10, 12 Submedian centromeres. No. 9 *may* have a secondary constriction in the proximal part of the long arm.

Group 13-15 (D)	Medium-sized chromosomes with nearly terminal centromeres (acrocentric chromosomes). Satellites have now been detected on all three chromosome pairs. The pairs are difficult to distinguish except by decreasing length (13 to 15).

(Continued on page 44)

43

Table 3-1. *(Cont.)*

Group 16-18 (E)	Rather short chromosomes.

> No. 16. Approximately median centromere. A secondary constriction has been frequently seen in the proximal part of the long arm.
> No. 17. Submedian centromere.
> No. 18. Submedian centromere. The shortest pair in the group. Short arm may be shorter than No. 17.

Group 19-20 (F)	Short chromosomes with approximately median centromeres. (They look like little "Xs".) They are distinctly smaller than E group, and the pairs cannot be distinguished.
Group 21-22 (G plus Y)	Very short chromosomes with nearly terminal centromeres (acrocentric chromosomes). Satellites have now been detected on both Nos. 21 and 22. The pairs cannot be distinguished.

> Y   The most common Y is larger than either 21 or 22; its centromere is often indistinct, and a secondary constriction is frequently seen in the long arm; the terminal region of the long arm may be poorly defined. Typically, the long arm chromatids appear to diverge less than those of other chromosomes.

Table 3-2.   Terminology of Centromere Position[5]

Centromere Position	Arm Ratio	Symbol	Other Term
Median sensu stricto	1.0	M	Metacentric
Median		m	Metacentric
	1.7		
Submedian		sm	Submetacentric
	3.0		
Subterminal		st	Subtelocentric
	7.0		
Terminal		t	Acrocentric*
Terminal sensu stricto	∞	T	Telocentric

* In common usage *acrocentric* is usually synonymous with *subterminal* or *subtelocentric*.

Table 3-3. Quantitative Characteristics of the Human Mitotic Chromosomes[2]

All measurements were made from cells of normal individuals, except those made by Fraccaro and Lindsten, which included cases of Turner syndrome. Column A is the relative length of each chromosome,* B is the arm ratio,† and C the centromere index.‡ as defined below.

	Tjio and Puck			Chu and Giles			Levan and Hsu			Fraccaro and Lindsten			Lejeune and Turpin			Buckton, Jacobs, and Harnden			Range		
	A	B	C	A	B	C	A	B	C	A	B	C	A	B	C	A	B	C	A	B	C
1	90	1·1	48	90	1·1	48	85	1·1	49	82	1·1	48	87	1·1	48	83	1·1	48	82–90	1·1	48–90
2	82	1·6	39	83	1·5	40	79	1·6	38	77	1·5	40	84	1·5	40	79	1·6	38	77–84	1·5–1·6	38–40
3	70	1·2	45	72	1·2	46	69	1·2	45	65	1·2	45	67	1·2	46	63	1·2	46	63–72	1·2	45–46
4	64	2·9	26	63	2·9	26	63	2·7	27	62	2·6	28	62	2·6	28	60	2·6	28	60–64	2·6–2·9	26–28
5	58	3·2	24	58	3·2	24	59	2·6	28	60	2·4	29	57	2·4	30	57	2·4	30	57–60	2·4–3·2	24–30
X	59	1·9	34	57	2·0	26	52	1·6	38	54	1·6	38	58	2·2	32	51	1·7	37	51–59	1·6–2·8	26–38
6	55	1·7	37	56	1·8	36	56	1·7	37	54	1·6	38	56	1·7	37	56	1·6	38	54–56	1·6–1·8	36–38
7	47	1·3	43	52	1·9	35	51	1·9	35	50	1·7	37	51	1·8	36	50	1·7	37	47–52	1·3–1·9	35–43
8	44	1·5	40	46	1·7	38	48	1·6	38	47	1·7	37	48	2·4	29	46	1·5	40	44–48	1·5–2·4	29–40
9	44	1·9	34	46	2·4	29	47	1·8	36	45	2·0	33	47	1·9	35	44	2·1	32	44–47	1·8–2·4	29–36
10	43	2·4	29	45	2·3	30	45	2·0	33	45	2·6	28	45	2·6	27	44	1·9	35	43–45	1·9–2·6	27–35
11	43	2·8	26	44	2·1	32	44	2·2	31	43	2·2	31	44	1·6	39	43	1·5	40	43–44	1·5–2·8	26–40
12	42	3·1	24	43	3·1	24	42	1·7	37	43	1·7	37	42	2·8	27	42	2·1	32	42–43	1·7–3·1	24–37
13	35	8·0	11	32	9·7	10	32	5·0	16	34	4·8	17	33	6·8	14	36	4·9	17	32–36	4·8–9·7	10–17
14	32	7·3	12	34	9·5	9	37	4·0	20	35	4·4	19	32	7·0	13	34	4·3	19	32–37	4·3–9·5	9–20
15	29	10·5	9	31	11·9	8	35	4·7	17	33	4·6	18	31	10·0	9	34	3·8	22	29–35	3·8–11·9	8–22
16	32	1·8	36	27	1·6	38	30	1·4	42	31	1·4	42	29	1·4	41	33	1·4	31	27–33	1·4–1·8	36–42
17	29	2·8	26	30	2·1	33	29	2·4	30	30	1·9	35	29	3·1	23	30	1·8	36	29–30	1·8–3·1	23–36
18	24	3·8	21	25	3·8	22	25	2·6	28	27	2·5	29	26	4·2	21	27	2·4	29	24–27	2·4–4·2	21–29
19	22	1·4	41	22	1·9	34	24	1·2	45	25	1·3	43	22	1·4	42	26	1·2	45	22–26	1·2–1·9	34–45
20	21	1·3	44	19	1·3	44	21	1·2	45	23	1·3	43	20	1·2	43	25	1·2	46	19–25	1·2–1·3	40–46
21	18	3·7	21	15	6·8	13	13	2·5	28	19	2·5	29	15	2·3	31	20	2·5	29	13–20	2·3–6·8	13–31
22	17	3·3	23	12	6·0	14	16	2·0	33	17	2·3	30	13	4·0	20	18	2·7	27	12–18	2·0–6·0	14–33
Y	19	∞	0	11	5·0	17	18	4·9	17	22	2·9	26	18	∞	0	18	4·9	17	11–22	2·9–∞	0–26

* Relative Length. The length of each chromosome relative to the total length of a normal, X-containing, haploid set, i.e., the sum of the lengths of the 22 autosomes and of the X chromosome, expressed per thousand.
† Arm Ratio. The length of the longer arm relative to the shorter one.
‡ Centromere Index. The ratio of the length of the shorter arm to the whole length of the chromosome.

45

Table 3-4. Nomenclature Symbols[2]

A-G	the chromosome groups
1-22	the autosome numbers (Denver System)
X, Y	the sex chromosomes
diagonal (/)	separates cell lines in describing mosaicism
plus sign (+) or minus sign (−)	when placed immediately after the *autosome number or group letter designation* indicates that the particular chromosome is extra or missing; when placed immediately after the *arm or structural designation* indicates that the particular arm or structure is larger or smaller than normal
question mark (?)	indicates questionable identification of chromosome or chromosome structure
asterisk (*)	designates a chromosome or chromosome structure explained in text or footnote
ace	acentric
cen	centromere
dic	dicentric
end	endoreduplication
h	secondary constriction or negatively staining region
i	isochromosome
inv	inversion
inv (p+q−) or inv (p−q+)	pericentric inversion
mar	marker chromosome
mat	maternal origin
p	short arm of chromosome
pat	paternal origin
q	long arm of chromosome
r	ring chromosome
s	satellite
t	translocation
tri	tricentric
repeated symbols	duplication of chromosome structure

Table 3-5. Recording of Numerical Aberrations[2]

1. The first item to be recorded is the total number of chromosomes, including the sex chromosomes, followed by a comma. The sex chromosome constitution is given next.

   Examples:

45, X	45 chromosomes, one X chromosome.
47, XXY	47 chromosomes, XXY sex chromosomes.
49, XXXXY	49 chromosomes, XXXXY sex chromosomes.

2. The autosomes are specified only when there is an abnormality present. If there is a numerical aberration of the autosomes, the group letter of the extra

46

or missing autosome and a plus (+) or minus (−) sign follows the sex chromosome designation.

Examples:

45, XX, C−	45 chromosomes, XX sex chromosomes, a missing C group chromosome.
48, XXY, G+	48 chromosomes, XXY sex chromosomes, an additional G group chromosome.

3. The plus or minus sign after a chromosome letter or number indicates that the entire autosome is extra or missing. When the extra or missing chromosome or chromosomes have been identified with certainty, the chromosome number may be used.

Examples:

45, XX, 16−	45 chromosomes, two X chromosomes, a missing No. 16 chromosome.
47, XY, 21+	47 chromosomes, XY sex chromosomes, an additional No. 21 chromosome.
46, XY, 18 +, 21 −	46 chromosomes, XY sex chromosomes, an extra No. 18 and missing No. 21.

4. A question mark (?) is used to indicate uncertainty. It may precede the group designation or in some cases the chromosome number.

Examples:

45, XX, ?C−	45 chromosomes, XX sex chromosomes, a missing chromosome which probably belongs in group C.
47,XX,?G+ 47,XX,G+ 47, XX, ?21+ 47,XX,21+	47 chromosomes, XX sex chromosomes, an additional small acrocentric chromosome recorded differently, depending on the amount of available information.

5. A triploid, polyploid, or aneuploid cell should be evident from the chromosome number and from the further designations.

Examples:

69,XXY
70, XXY, G+

6. An endoreduplicated metaphase can be indicated by preceding the karyotype designation with the abbreviation *end.* If multiplicity of endoreduplications is to be indicated, an Arabic numeral can be used before *end* to indicate the number of endoreduplicated cells.

Examples:

end46,XX
2end46,XX or 4end46,XX

7. The chromosome constitution of the different cell lines in chromosome mosaics are listed in numerical or alphabetical order, irrespective of the frequencies of the cell types in the individual studied. The karyotype designations are separated by a diagonal (/).

Examples:

45,X/46,XY	A chromosome mosaic with two cell types, one with 45 chromosomes and a single X, the other with 46 chromosomes and XY sex chromosomes.
46,XX/46,XY	A chromosome mosaic with both XX and XY cell lines.
46,XY/47, XY,G+	A chromosome mosaic with a normal male cell line and a cell line with an extra G group chromosome.
45,X/46, XX/47,XXX	A triple cell line mosaic.

Table 3-6. Recording of Structural Alterations[2]

1. The short arm of a chromosome is designated by the lowercase letter "p," the long arm by the letter "q," a satellite by the letter "s," a secondary constriction by the letter "h," and the centromere by the abbreviation "cen." The symbol is placed after the chromosome designation.

2. Increase in length of a chromosome arm is indicated by placing a plus sign (+) and a decrease in length by placing a minus sign (−) after the arm designation.

   Examples:

2p+	Increase in length of the short arm of No. 2.
Bp−	Decrease in length of the short arm of a B group chromosome.
Gq−	Decrease in length of the long arm of a G group chromosome.

3. When one arm of a mediocentric chromosome (Nos. 1, 3, 19, 20) is changed, this is indicated by placing a question mark between the chromosome designation and the plus or minus sign.

   Example:

3?+	Increase in length of one arm of a No. 3.

4. The result of a pericentric inversion is indicated by p+q− or p−q+, which is enclosed in parentheses and preceded by the abbreviation "inv."

   Example:

inv(Dp+q−)	A pericentric inversion involving the long and short arms of a D group chromosome.

5. A translocation is indicated by the letter "t" followed by parentheses which include the chromosome involved. The separation of the chromosomes within the parentheses by a semicolon (;) indicates that two structurally altered chromosomes are present and that the translocation is balanced.

   Example:

46,XY,t(Bp−;Dq+) or 46,XY,t(Bp+;Dq−)	A balanced reciprocal translocation between the short arm of a B and the long arm of a D group chromosome.

6. Translocations involving a sex chromosome and an autosome should be designated in the same way. The remaining normal sex chromosome is written in its usual position after the chromosome number, and the other sex chromosome which is involved in the translocation is included in parentheses preceding the autosome concerned.

   Examples:

46,X,t(Xq+; 16p−)	A reciprocal translocation between the long arm of an X and the short arm of a No. 16 in a female.
46,Y,t(Xq+; 16p−)	The same translocation in a male.
46,X,t(Yp+; 16p−)	A reciprocal translocation between the short arm of a Y and the short arm of a No. 16.

7. Where a "centric fusion" type of translocation results in duplication of part of one of the chromosomes involved, this could be written as follows:

   Example:

46,XX,D−,t(DqGq)+	46 chromosomes, XX sex chromosomes, one chromosome missing from the D group. The long arm of this chromosome is united with the long arm of a G group chromosome. Since there are four normal G group chromosomes, part of a G is present in triplicate.

8. When family studies clearly show that a particular chromosome has been inherited from the mother or the father, this fact may be indicated by the abbreviations "mat" or "pat."

   Examples:

46,XY,t(Bp—;Dq+)	46 chromosomes, XY sex chromosomes, balanced reciprocal translocation between the short arm of a B and the long arm of a D group chromosome.
46,XY,Bp—pat *or* 46,XY,Dq+pat	The son of the above father has inherited only one of the two abnormal chromosomes, depending on which abnormal chromosome has been transmitted.
46,XY,t(Bp—; Dq+) pat	The son has inherited both chromosomes involved in the translocation.

9. Duplicated chromosome structures are indicated by repeating the appropriate designation. If the structures are abnormal or appear in an unusual place these should be indicated.

   Examples:

46,XX,Gpss	46 chromosomes, XX sex chromosomes, one G group chromosome with double satellites on the short arm.
46,XY,18ps	46 chromosomes, XY sex chromosomes, a No. 18 with satellited short arms.
46,XX,Gpsqs	46 chromosomes, XX sex chromosomes, a G group chromosome with both long and short arms satellited.
Gs+	Enlarged satellites on a G group chromosome.

10. Isochromosomes are designated by the lower-case letter "i" placed after the chromosome arm involved. A question mark is used to designate uncertainty.

    Examples:

46,XXqi	Isochromosome for the long arm of one X chromosome.
46,XXq?i	Presumptive isochromosome for the long arm of one X chromosome.

11. Ring chromosomes are indicated by the letter "r" placed after the chromosome involved.

    Examples:

46,XXr	One ring X chromosome.
46,XY,Br	One ring B chromosome.

12. In describing cells damaged by ionizing radiation, chemicals, viruses, etc., the system of nomenclature that has been described should be used where applicable. This may not be possible if the cell contains a grossly unbalanced chromosome complement. In such instances the following convention is suggested. The chromosome count in a given cell should include all centric chromosome structures present in that cell regardless of the number of centromeres. Unidentified chromosomes are indicated by "mar" (marker). Acentric fragments are not included in the count but may be indicated by "ace." Dicentric and tricentric chromosomes are counted as one body and indicated by "dic" and "tri." If necessary, an asterisk (*) following a chromosome designation may be used to draw attention to an explanation in the text.

    Example:

48,?X?X,F—,dic+,mar1+,mar2+,ace*	A cell derived from a normal female with a total of 48 centric chromosome structures, one missing F group chromosome, a dicentric and an acentric fragment, as well as two unidentified marker chromosomes.

Table 3-7a. Terminal Labeling Patterns of Normal Human Chromosomes, as Defined by Tritiated Thymidine Autoradiography.[2]

Chromosome	Finish Labeling Earlier	Finish Labeling Later	Major Reference
No. 1	Distal segment one arm (short)	Proximal portion one arm (long) with secondary constriction	Compared to the whole chromosome complement
No. 4		Long arm	Compared to each other
No. 5	Long arm		
No. 13 (D_1)		Middle or lower half of long arm	
No. 14 (D_2)		Near centromere	Compared to each other
No. 15 (D_3)	Entire chromosome		
No. 17	Entire chromosome		Compared to each other
No. 18		Entire chromosome	
No. 19	Entire chromosome		Compared to the whole chromosome complement
No. 20	Entire chromosome		
All Xs in excess of one		Entire(?) chromosome	Compared to the whole chromosome complement
Y		Entire chromosome	Compared to other G group chromosomes

Table 3-7b. Terminal Labeling Patterns of Abnormal Human Chromosomes, as Defined by Tritiated Thymidine Autoradiography.

1. No. 4 is later to replicate over the long arm than No. 5 and is less commonly deleted in the B chromosome deletion syndrome than is No. 5.[11]
2. No. 13 (D_1) is later to replicate over the distal long arm than are Nos. 14 (D_2) and 15 (D_3) and is commonly trisomic in D trisomy syndrome.[9]
3. No. 21 (G_1) is found, by some investigators, to replicate later than No. 22 and is trisomic in Down's syndrome (mongolism).[10] However, other investigators report difficulty in distinguishing No. 21 and No. 22 by their terminal labeling patterns.
4. As a rule, structurally abnormal X chromosomes (rings, isochromosomes, or deleted) are late to replicate as compared to the entire chromosome complement.[2]
5. The most common D/G translocation chromosome associated with Down's syndrome is 14/21. Less commonly it is 15/21.[4]

50

REFERENCES

1. Bartalos, M., and Baramki, T. A.: *Medical Cytogenetics.* Baltimore, Williams & Wilkins Co., 1967.

2. Chicago Conference: Standardization in Human Cytogenetics. Birth Defects: Original Article Series, *II*:2, New York, The National Foundation, 1966.

3. Denver Conference: A Proposed System of Nomenclature of Human Mitotic Chromosomes. Lancet, *i:*1063, 1960.

4. Hecht, F. H., et al.: Nonrandomness of Translocations in Man. Preferential Entry of Chromosomes into 13-15/21 Translocations. Science, *161:*371, 1968.

5. Levan, A., Fredga, K., and Sandberg, A. A.: Nomenclature for Centromeric Position on Chromosomes. Hereditas, *52:*201, 1964.

6. London Conference: On the Normal Karyotype. Cytogenetics, *2:*264, 1963.

7. Swanson, C. P., Merz, T., and Young, W. J.: *Cytogenetics.* Englewood Cliffs, New Jersey, Prentice-Hall, Inc., 1966.

8. Yunis, J. J., ed.: *Human Chromosome Methodology.* New York, Academic Press, 1965.

9. Yunis, J. J., Hook, E. B., and Mayer, M.: Deoxyribonucleic-Acid Replication Pattern of Trisomy D_1. Lancet, *ii:*935, 1964.

10. Yunis, J. J., Hook, E. B., and Mayer, M.: D.N.A. Replication Analysis in Identifying the Cytogenetic Defect in Down's Syndrome (Mongolism). Lancet, *i:*465, 1965.

11. Warburton, D., Miller, D. A., Miller, O. J., Breg, W. R., de Capoa, A., and Shaw, M. W.: Distinction Between Chromosome 4 and Chromosome 5 by Replication Pattern and Length of Long and Short Arms. Amer. J. Hum. Genet., *19:*399, 1967.

4

Analysis of Chromosomes
from Peripheral Blood

Techniques using a *mitogenic* agent to stimulate division of peripheral blood leukocytes, involve *short-term tissue culture.* Standard tissue culture procedures are required, but the culture is short-lived. Initially there was some question as to what type of blood cell participated in the "burst" of mitosis or *blastoid reaction.* It is now agreed that small lymphocytes are involved.[5] They cannot be induced to undergo another wave of divisions following addition of more mitogenic agent. The maximal response usually occurs by the third day,[2] and occasional mitoses are still seen for up to two weeks. The mechanism of blastoid reaction and the fine structural and molecular events in the cell are matters of intensive investigation.[6, 7] Although phytohemagglutinin, an extract from red kidney beans, is the most commonly used mitogenic agent, the blastoid reaction can be produced by a variety of other agents, including dextran,[1] and various antigenic substances.[5] The simple mixing of different types of cells can produce a burst of mitosis.[3] To understand what is going on, purification of the mitogenic substance or substances is a necessary step.[4]

52

MACROMETHOD FOR SHORT-TERM TISSUE CULTURE OF HUMAN PERIPHERAL BLOOD

Equipment and Reagents

1. 10-ml plastic disposable syringe with No. 20 needle, sterile.
2. Sodium heparin, aqueous solution, 1000 USP units/ml, sterile.
3. Phytohemagglutinin (a dried preparation is available from Burroughs Wellcome & Co., London, England). Reconstitute with sterile triple distilled water. An extract of red kidney beans may be prepared, but different lots will vary in mitogenic activity (see protocol in this chapter).
4. Tissue culture vessel. 2-oz prescription bottles which are made of soft glass and are disposable are very satisfactory. They should be rinsed with triple distilled water prior to use and autoclaved (see Chapter 12 on sterilization for tissue culture). Siliconization is not necessary. If a gastight cap is needed, the caps which come with the bottles may be lined with #22 Teflon

liners (Arthur H. Thomas Co., Philadelphia, Pa.). Disposable plastic tissue culture containers are also available (Falcon Plastics, Los Angeles, Calif.).

5. Tissue culture medium. Several types of defined medium for mammalian cells can be made from standard protocols (see protocol for Dulbecco and Vogt's medium [D & V] in Chapter 12), or are available commercially (Hyland Lab., Los Angeles, Calif.; Flow Lab., Rockville, Md.; Microbiological Assoc., Bethesda, Md.). Puck's F-10 or F-12, Dulbecco and Vogt's medium, Medium 199, Waymouth's medium, NCTC-109 are all satisfactory. Serum must be added and is available commercially from the companies that supply medium. A concentration of 15% fetal calf serum (Flow Lab., Rockville, Md.) is recommended. The medium should contain phenol red indicator. Antibiotics are optional.

Procedure

1. Wet 10-ml syringe with heparin. Draw 10 ml blood, clear needle, and stand syringe on plunger for $\frac{1}{2}$-2 hr. When plasma has cleared of RBCs, bend needle and extrude plasma into sterile tube.

or

Draw 10 ml blood in plain syringe. Add to tube containing 0.2 to 0.5 ml of heparin. Allow RBCs to settle out. Aspirate plasma from RBCs with sterile Pasteur pipette.

2. A leukocyte count *may* be made on plasma, using a standard procedure for WBC count, in a hemocytometer. Total cell count should be at least 2×10^6/ml.

3. Add 0.5 to 1.0 ml plasma to culture medium in 2-oz culture container. Total volume should not exceed about 10 ml.

4. Add 0.1 to 0.2 ml phytohemagglutinin, shake culture, and incubate at 37°C. Gastight containers should be gassed with CO_2 (10% for D & V medium, 5% for most others). If gas exchange is allowed, the containers should be placed in a humidified, gas-flow incubator with the appropriate concentration of CO_2 (5% or 10%). It is very important to maintain the cultures at the correct pH. The color of phenol red indicator should be pink-orange (pH 7.4).

5. During incubation for about 3 days (72 hr) the cultures may be shaken 3 times daily.

References: Moorhead, P. S., Nowell, P. C., Mellman, W. J., Battips, D. M., and Hungerford, D. A.: Chromosome Preparations of Leukocytes Cultured from Human Peripheral Blood. Exp. Cell Res., *20*:613, 1960.

Mellman, W. J.: Human Peripheral Blood Leucocyte Cultures. *In: Human Chromosome Methodology,* J. J. Yunis, ed. New York, Academic Press, 1965.

PREPARATION OF CHROMOSOMES FROM HUMAN BLOOD CULTURES

Equipment and Reagents

1. Colchicine 50.0 μgm/ml in sterile triple distilled H_2O.

    or

    Vinblastine sulfate (Velban) 10.0 μgm/ml in sterile triple distilled H_2O.
2. Heavy duty, graduated 12-ml centrifuge tubes.
3. Centrifuge with rpm indicator.
4. Distilled H_2O.
5. Disposable Pasteur pipettes, 1-ml rubber bulbs.
6. Isotonic, buffered salt solution (see protocol in Chapter 12 for phosphate buffered saline, PBS).
7. One part glacial acetic acid: 3 parts absolute methanol fixative, freshly made.
8. 45% acetic acid.
9. Microscope slides precleaned by soaking in 95% alcohol. Dry with lint-free towel or Miracloth (Miracloth Sales, Milltown, N. J.).
10. Giemsa stain, freshly diluted. Reconstitute from commercial stock solution by adding 2 ml stock stain, 2 ml phosphate buffer pH 6.4, to make 50 ml in distilled H_2O. (See Chapter 7 on chromosome stains.)

Procedure

1. After peripheral blood cultures have incubated approximately 72 hr, add sterile colchicine to a final concentration of 0.5 μgm/ml* or sterile Velban to a final concentration of 0.1 μgm/ml.
2. Continue incubation for 3 more hr (up to 5 hr with colchicine).

---

* 0.1 μgm/ml final concentration of colchicine may be preferable.

3. Pour medium and cells into centrifuge tube. Centrifuge at 600 to 800 rpm for 6 to 8 min.
4. Pour off supernatant. Add several ml PBS; suspend cells gently by pushing out with Pasteur pipette.
5. Centrifuge as before.
6. Remove supernatant, leaving 0.5 ml.
7. Fill to 2.0 mark with distilled H_2O.
8. Suspend cells gently as before.
9. Incubate at 37°C for 10 min.
10. Shake gently and centrifuge as before.
11. Aspirate all supernatant as close to top of button as possible.
12. Add 1 to 2 ml 1:3 fixative without disturbing button of cells.
13. Allow to stand at room temperature for 30 min.
14. Resuspend cells by pushing out with Pasteur pipette.
15. Centrifuge as before.
16. Aspirate all supernatant.
*17. Add about 0.2 ml 45% acetic acid to make a dense cell suspension. Add less for a very small button.
18. Prepare one slide by allowing a small drip to flow down slide. Dry quickly. Metaphase spread may be checked without stain if the microscope light is turned low. Prepare remainder of slides if spread is good. Otherwise, leave cells longer in 45% acetic acid.
19. Stain about 5 min in Giemsa. Rinse in distilled H_2O. Air dry.

References: Darlington, C. D., and La Cour, L. F.: *The Handling of Chromosomes.* London, George Allen & Unwin Ltd., 1960.
Mellman, W. J.: Human Peripheral Blood Leucocyte Cultures. In: *Human Chromosome Methodology,* J. J. Yunis, ed. New York, Academic Press, 1965.

MICROMETHOD FOR SHORT-TERM CULTURE OF HUMAN PERIPHERAL BLOOD

Procedure

1. Prepare culture medium plus 15% fetal calf serum and container as described for macromethod.

---

* If flame-dry procedure is used for slides, omit 45% acetic acid and add two changes of fresh 1:3 fixative, drop on chilled slide wet with distilled H_2O, and dry fixative by passing slide through a flame briefly. (See Chapter 7 for flame-dry technique and also for chromosome squash technique.)

2. Add about 0.4 ml heparin to about 10 ml medium in 2-oz container.
3. Add 0.1 to 0.2 ml phytohemagglutinin.
4. Allow 3 to 4 drops of blood to drip from a finger prick or a heel stab into culture container, which is shaken gently to prevent clotting.
5. Proceed as for macromethod, with the modification that the red cells are hemolyzed during hypotonic treatment and go into the pellet after centrifugation. Therefore, more changes of 1:3 fixative are indicated.

or

1. Obtain 0.1 to 0.2 ml sterile capillary or venous blood in a 1-ml plastic syringe wet with heparin.
2. Inoculate blood directly into medium in 2-oz container.
3. Add 0.1 to 0.2 ml phytohemagglutinin.
4. Proceed as for macromethod but with additional changes of 1:3 fixative.

References: Bishum, N. P., Morton, W. R. N., and Froggat, P.: Assessment of Two Macromethods and Three Micromethods of Culturing Human White Blood Cells for Chromosome Analysis. J. Med. Genet., *2:*251, 1965.

Edwards, J. H.: Chromosome Analysis from Capillary Blood. Cytogenetics, *1:*90, 1962.

Huang, S. W., and Emanuel, I.: Microtechnique for Culturing Leucocytes for Improved Spread of Chromosomes. Lancet, *i:*707, 1965.

Hungerford, D. A.: Leukocytes Cultured from Small Inocula of Whole Blood and the Preparation of Metaphase Chromosomes by Treatment with Hypotonic KCl. Stain Techn., *40:*333, 1965.

Timson, J.: A Simple Rapid Chromosome Micromethod. Lancet, *i:*50, 1967.

MAIL-IN METHOD FOR HUMAN PERIPHERAL BLOOD CULTURE

References: Anders, J. M., Moores, E. C., and Emanuel, R.: Chromosome Preparation from Leukocyte Culture. A Simplified Method for Collecting Samples by Post. J. Med. Genet., *3:*74, 1966.

Birdwell, T. R., Eggen, R. R., and Dimmette, R. M.: A Simple Mail-In Method for Chromosome Analysis. Amer. J. Clin. Path., *45:*153, 1966.

KIT METHOD FOR HUMAN PERIPHERAL BLOOD CULTURE

Tissue culture kits for the processing of human peripheral blood are available commercially (Difco Lab., Detroit, Mich.; Hyland Lab., Los Angeles, Calif.; Grand Island Biological Co., Grand Island, N. Y.). These kits are recommended if chromosome analyses are performed at infrequent intervals or if tissue culture is not a routine procedure in the laboratory. Kits are also useful for persons just learning cytogenetic procedures, especially in laboratories in which routines are not yet established.

HUMAN LEUKEMIC PERIPHERAL BLOOD CULTURE

Peripheral blood from patients with various types of leukemia may require modifications in the culturing procedure to obtain the best results. Differing results are reported with different culture media. Various types of modifications have been suggested:

1. Be sure that the final serum concentration (fetal calf plus patient's own serum) in the culture medium is about 30%.
2. Start with a high initial inoculation of leukocytes (2 to 3 $\times$ 10^6/ml final concentration). Draw 20 ml blood initially.
3. Introduce cell suspension into cold medium to prevent clumping.
4. Incubate at 38°C instead of 37°C for about 48 hr instead of about 72 hr.
5. After 72-hr incubation (4th day) change the medium but do not add more phytohemagglutinin. Harvest cultures on the 6th day.
6. Follow the usual incubation temperature but vary the length of time after application of phytohemagglutinin and before harvesting of chromosomes (2-day incubation following introduction of phytohemagglutinin is usual).

References: Moore, W., Gillespie, L. J., and Dolimpio, D. A.: Cultivating Leukemic Lymphocytes. Lancet, *i:*363, 1968.

Moorhead, P. S., Nowell, P. C., Mellman, W. J., Battips, D. M., and Hungerford, D. A.: Chromosome Preparations of Leukoctyes Cultured from Human Peripheral Blood. Exp. Cell Res., *20:*613, 1960.

Rozynkowa, D.: Cultivating Leukaemic Lymphocytes. Lancet, *i:* 1255, 1968.

SHORT-TERM TISSUE CULTURE AND PREPARATION OF CHROMOSOMES FROM PERIPHERAL BLOOD OF OTHER MAMMALS

The general rules for short-term culture of human peripheral blood are also applicable to other mammals. However, variations in some of the details of the procedure may give better results for each particular species to be studied. It is well recognized, for instance, that phytohemagglutinin or a particular lot of phytohemagglutinin may not be equally effective for all species. Therefore it is recommended that publications covering the particular mammal be consulted or the laboratory with the most experience be contacted.

Reference: Hsu, T. C., and Benirschke, K.: *An Atlas of Mammalian Chromosomes.* New York, Springer-Verlag, 1967.

SHORT-TERM CULTURE OF AUTOPSY SPLEEN LEUKOCYTES (OR INFANT THYMUS)

1. Specimen must be obtained by sterile technique.
2. Rinse with buffered, isotonic salt solution. (See protocol in Chapter 12 for PBS.)
3. Dissect out pulp of spleen, mince, and rinse again. Allow pieces of spleen to settle out and obtain a supernatant containing white cells and some red cells.
4. Proceed as for peripheral blood culture.

Reference: Conen, P. E., and Erkman, B.: Necropsy Spleen Samples for Chromosome Cultures. Lancet, *i:*665, 1964.

SHORT-TERM CULTURE OF POSTMORTEM LEUKOCYTES

Reference: Macek, M.: A Simple Method for Postmortem Cultivation of Human Leukocytes. Cytologia, *32:*165, 1967.

SHORT-TERM CULTURE OF HUMAN THORACIC DUCT LEUKOCYTES

Reference: Lindahl-Kiessling, K., Werner, B., and Book, J. A.: Short Term Cultivation of Human Thoracic Duct Lymphocytes with *Phaseolus Vulgaris* Extract. Hereditas (Lund), *53:*40, 1965.

59

SHORT-TERM CULTURE OF FETAL ASCITIC FLUID

Reference: Chang, T. D., and Bowman, J. M.: Chromosomes from Foetal Ascitic Fluid. Lancet, *i:*1431, 1968.

LONG-TERM CULTURE OF HUMAN LEUKOCYTES

References: Armstrong, D.: Serial Cultivation of Human Leukemic Cells. Proc. Soc. Exp. Biol. Med., *122:*475, 1966.

Clarkson, B., Strife, A., and de Harven, E.: Continuous Culture of Seven New Cell Lines (SK-L1 to 7) from Patients with Acute Leukemia. Cancer, *20:*926, 1967.

Miles, C. P., O'Neill, F., Armstrong, D., Clarkson, B., and Keane, J.: Chromosome Patterns of Human Leukocyte Established Cell Lines. Cancer Res., *28:*481, 1968.

Moore, G. E., Ito, E., Ulrich, K., and Sandberg, A. A.: Culture of Human Leukemia Cells. Cancer, *19:*713, 1966.

Moore, G. E., Gerner, R. E., and Franklin, H. A.: Culture of Normal Human Leukocytes. J.A.M.A., *199:*519, 1967.

PROTOCOL—RED KIDNEY BEAN (RKB) PHYTOHEMAGGLUTININ

Procedure

1. Wash 20 gm red kidney beans (*Phaseolus vulgaris*) with H_2O.
2. Soak overnight in refrigerator ($4°C$).
3. Decant excess H_2O and place beans in homogenizer or blender.
4. Add 30 ml 0.85% (to 0.9%) NaCl and grind beans to a fine pulp.
5. Add 70 ml additional 0.85% NaCl.
6. Allow to extract for 24 hr in refrigerator, mixing occasionally.
7. Centrifuge at high speed for 1 hr.
8. Pipette off supernatant, dispense, and freeze.
9. Thaw concentrate, dilute 1:25 with 0.85% NaCl, and sterilize by pressure filtration. Freeze sterilized material in small aliquots.
10. Mitogenetic activity varies and should be tested for each extraction.
11. Use 0.2 ml per 10 ml of culture.

Reference: Rigas, D. A., and Osgood, E. E.: Purification and Properties of the Phytohemagglutinin of *Phaseolus vulgaris*. J. Biol. Chem., *212:*607, 1955.

REFERENCES

1. Hutton, W. E., and Watson, E. D.: Dextran-Glucose and the Cultivation of Leukocytes. Proc. Soc. Exp. Biol. Med., *116:*924, 1964.

2. Nowell, P. C.: Phytohemagglutinin: An Initiator of Mitosis in Cultures of Normal Human Leukocytes. Cancer Res., *20:*462, 1960.

3. Priest, J. H.: Personal observation.

4. Rivera, A., and Mueller, G. C.: Differentiation of the Biological Activities of Phytohemagglutinin. Nature, *212:*1207, 1966.

5. Robbins, J. H.: Tissue Culture Studies of the Human Lymphocyte. Science, *146:*1648, 1964.

6. Rubin, A. D., and Cooper, H. L.: Evolving Patterns of RNA Metabolism during Transition from Resting State to Active Growth in Lymphocytes Stimulated by Phytohemagglutinin. Proc. Nat. Acad. Sci., U.S.A., *54:*469, 1965.

7. Salzman, N. P., Pellegrino, M., and Franceschini, P.: Biochemical Changes in Phytohemagglutinin Stimulated Human Lymphocytes. Exp. Cell Res., *44:* 73, 1966.

5

Analysis of Chromosomes from Cells in Serial Tissue Culture

Cells may be grown in *long-term culture* for chromosome analysis. This method has these advantages: (1) multiple preparations may be made without returning to the patient; (2) cells may be stored in liquid nitrogen and retrieved when needed; (3) a satisfactory sampling of metaphases is uniformly obtained if the culture is growing well; and (4) the effects of various environmental agents and manipulations may be studied. Chapter 12 deals with tissue culture techniques and covers many details of long-term tissue culture. An outline of the handling of cells for chromosome preparations is included in this chapter.

> *Reference:* Harnden, D. G., and Brunton, S.: The Skin Culture Technique. In: *Human Chromosome Methodology*, J. J. Yunis, ed. New York, Academic Press, 1965.

LONG-TERM TISSUE CULTURE OF HUMAN CELLS

Procedure—Cells in Culture Bottles

1. Assemble medium, 0.25% trypsin solution, and canisters of Pasteur pipettes, 5- and 10-ml serological pipettes.

62

2. Take a confluent 8-oz prescription bottle to be subcultured and decant medium. With 5-ml pipette add about 5 ml PBS or other balanced salt solution (see Chapter 12). Put pipette aside. Rotate the bottle so that top and sides, as well as bottom, are washed. Decant fluid.
3. Repeat washing with PBS.
4. With Pasteur pipette and 1-ml sterile rubber bulb, add 1 ml of trypsin and discard Pasteur pipette. Make sure trypsin comes in contact with entire bottom surface of bottle. Allow to stand at room temperature for 1 min; then decant trypsin. Cap and incubate at 37°C for 3 to 5 min. Check with inverted microscope to insure that cells are removed from bottom surface.*
5. Add 9 ml of D & V or other medium (see Chapter 12) to each of 2 new labeled bottles (1:2 split) or to each of 4 new labeled bottles (1:4 split), using the same 10-ml serological pipette. Set pipette aside. Make sure medium is pink-orange in color. If it is bright pink, run CO_2 (10% for D & V, 5% for most

---

* If cells are not removed, continue incubation for 5 min and recheck. If cells remain attached, add 1 ml trypsin and reincubate 5 plus min.

other media) into gas phase above medium until color is pink-orange.

6. With 5-ml pipette add medium to cells to a total of 2 ml if subdivision is 2:1, or 4 ml of medium if 4:1. With same pipette suspend cells well, and pipette 1 ml into each bottle set up in step 5 above. Cap loosely and place in 5% or 10% CO_2 incubator. Alternatively, run 5% or 10% CO_2 into bottle for about 15 sec, being sure to direct gas flow in gas phase of bottle, not into medium; cap tightly and place in nongas flow incubator.

General

1. When fluid is decanted from bottle, the mouth of the bottle should be flamed when the cap is removed and after decantation.

2. Fluid decanted from a bottle should be poured down funnel into flask, and funnel should subsequently be washed down with 1:10 dilution of bleach (Clorox) from a wash bottle.

3. A pipette that has been introduced into a bottle containing cells should never be returned to stock bottles of medium, PBS, or trypsin.

4. Never subculture all of your cells on the same day. Set up a routine for subculturing on alternate days, using red-coded medium, PBS, and trypsin on odd days and blue on even days.

5. Blow out cotton plugs from each serological pipette as soon as possible after using, and place tip up in soaking solution. Used pipettes should not be allowed to dry out.

6. Use your own bottle of medium, trypsin, and PBS. Initial. Never use anyone else's.

7. When you return a bottle of medium to the refrigerator for reuse, always make sure the cap is tight. Check color of medium before you put it on cells. If pH is too alkaline (bright pink color), medium will need to be gassed with a sterile pipette (10% CO_2 for D & V, 5% for most other media) for 1 min or until color is right. Do not run gas flow directly into medium, or there will be too much foaming.

8. It is unwise to economize by retaining an old bottle to use as one of the new bottles. The main problems with this method are

(a) accumulation of debris and old medium, and (b) splitting of cells is not exact.

9. For routine cell maintenance, at least 2 new bottles should always be started. Then on the next subculture, one can be subcultured and the other held in reserve, in case of contamination or growth problems. It is usually wise to keep in reserve several of these older bottles dating back for more than 1 subculture. To keep bottles from more than 3 different old subculturings is usually unnecessary and wastes incubator space.

Procedure—Cells in Roller Bottles

Monolayer cells can be grown in large quantities in roller bottles. The apparatus to roll the bottles at controlled and variable speeds and permanent or disposable bottles are available from Bellco Biological Glassware and Equipment, Vineland, N. J. Procedures for culture differ in quantity but not in quality from those for standard culture bottles. General suggestions for human diploid cells are listed here.

1. Initial cell inoculation should not be less than 2×10^6 cells for a bottle of 840 cm² growth area (the smallest size). Higher cell inoculations or repeated inoculations will result in more rapid confluency.

2. Cells grow well in volumes of medium not less than 50 ml per 840 cm² bottle. Volumes between 50 ml and 150 ml are usual. Medium changes are indicated as the medium becomes acid.

3. Bottles may be gassed to the appropriate CO_2 concentration. Caps are gastight.

4. Bottles should be rolled at the lowest speed. A more rapid speed is used for trypsinization and for about 1 hr following inoculation of new cells or medium change.

5. Rinses with balanced salt solution prior to trypsinization should be no less than 100 ml volume. Trypsin (0.25%) is added in quantities of about 5 ml. Cell dispersion takes approximately the same time as for standard bottles. It is essential to roll the bottles fairly rapidly during trypsinization.

6. The usual yield for human diploid cells at confluency is not less than 2×10^7 cells per bottle of 840 cm² growth area.

65

PREPARATION OF CHROMOSOMES FROM HUMAN
CELLS IN LONG-TERM CULTURE

Equipment and Reagents

1. Colchicine 50.0 μgm/ml in sterile triple distilled H_2O.

   or

   Vinblastine sulfate (Velban) 10.0 μgm/ml in sterile triple distilled H_2O.
2. Heavy duty, graduated 12-ml centrifuge tubes.
3. Centrifuge with rpm indicator.
4. Distilled H_2O.
5. Disposable Pasteur pipettes, 1-ml rubber bulbs.
6. Isotonic buffered salt solution (PBS), sterile and nonsterile.
7. One part glacial acetic acid:3 parts absolute methanol fixative, freshly made.
8. 45% acetic acid.
9. Microscope slides precleaned by soaking in 95% alcohol. Dry with lint-free towel or Miracloth.
10. Giemsa stain, freshly diluted. Reconstitute from commercial stock solution by adding 2 ml stock stain, 2 ml phosphate buffer pH 6.4, to make 50 ml in distilled H_2O. (Also see Chapter 7 on chromosome stains.)
11. A solution of 0.25% trypsin and 0.25% Versene in balanced salt solution (Chapter 12, solutions to disperse cells).

Procedure—Cells in Tissue Culture Bottles

Select a bottle of cells which are nearly confluent. They should not be overly confluent or overly sparse, since in either case the number of cells in mitosis is likely to be low.

1. Sterile technique should be used on the cells through step 6. Add 0.5 μgm/ml (or 0.1 μgm/ml) colchicine final concentration. Continue incubation at 37°C for 3 to 5 hr.

   or

   Add 0.1 μgm/ml vinblastine sulfate (Velban). Continue incubation at 37°C for 1 to 3 hr.
2. Pour off medium and save in heavy duty graduated centrifuge tube. (It is wise to save medium and rinses because metaphases may be lost in the fluids over the cell monolayer before it is trypsinized.)

3. Centrifuge medium at 600 to 800 rpm for about 8 min. Pour off supernatant.

4. Add about 4 ml balanced solution (PBS) to 8-oz prescription bottle of cells, rotate, pour off into same centrifuge tube. Repeat rinse procedure and add to centrifuge tube.

5. Add 1 ml 0.25% Versene-0.25% trypsin and incubate 1-plus min until cells come off.

6. Add about 2 ml PBS, suspend cells well, hit bottle with hand to dislodge cells further. Add cell suspension to the same centrifuge tube.

7. Centrifuge as before.

8. Pour off all supernatant. Add to 0.5 ml mark with PBS. Add to 2.0 ml mark with distilled H_2O.

9. Make sure cells are well suspended (by pushing out gently with Pasteur pipette) and allow to sit at room temperature for 10 min.

10. Centrifuge as before.

11. Add about 3 ml fresh fixative (1 part glacial acetic acid to 3 parts absolute methanol) down side of tube without disturbing button of cells.

12. Allow to stand at room temperature for 30 min. Suspend cells and centrifuge as before. Remove all of supernatant.

*13. Add 0.1 ml (or less) 45% acetic acid (depending on size of button). Allow to stand 5 min.

14. Prepare a test slide by allowing a drip of cell suspension to run down a slide precleaned with 95% alcohol. Check chromosome spread on a microscope with the light turned down.

15. If spread is good, prepare remainder of slides. Otherwise, make another test slide after about 20 min in 45% acetic acid (or longer).

16. Stain about 5 min in Giemsa stain. Rinse in distilled H_2O. Air dry.

17. Mounting is optional.

---

* Alternative methods for preparing slides: Add fresh 1:3 fixative instead of 45% acetic acid; centrifuge; remove supernatant. Repeat. Or add a small amount of 1:3 fixative and prepare flame-dried slides or make squash preparations (see Chapter 7).

Procedure—Cells in Tissue Culture Petri Dishes

1. The same steps are followed, but volumes of rinses, etc., should be adjusted to the size of the container. Swirl dish to rinse.
2. When cells are dispersed with 0.25% trypsin-0.25% Versene, add PBS and suspend cells gently with Pasteur pipette.

PREPARATION OF CHROMOSOMES FROM HUMAN CELLS GROWING ON COVERGLASSES

The usual method for examination of chromosomes in cultured monolayer cells is to suspend them and process the suspended cells. Human fibroblast-like cells, however, may be processed as monolayers if care is taken not to dislodge mitoses that tend to be less firmly attached than are interphases. The chromosomes are well spread. A modification of the method of Tjio is described here.

1. Prepare the desired number of 60-mm tissue culture Petri dishes (at least 6), each containing a sterile 22 × 40 mm, No. 1 thickness coverglass, precleaned in 95% alcohol. It is not necessary to anchor the coverglass, but the dishes should be handled as carefully as possible.
2. Inoculate 3×10^5 cells (approximately) in 4 ml in each dish, according to the replicate plating protocol described in Chapter 15. This procedure involves trypsinization of a rapidly growing monolayer culture, cell count, suspension of the total number of cells needed for all dishes in the total amount of medium needed for all dishes, and inoculation of 4 ml of cell suspension to each dish.
3. Incubate the dishes at 37°C in proper CO_2 concentration for about 48 hr.
4. Without interruption of the incubation, add colchicine to a final concentration of 0.1 μgm/ml (up to 0.5 μgm/ml). Continue incubation for another hr.
5. Pour off the medium and rinse the dish once with balanced salt solution. Flood with 0.17% saline at 37°C. After 5 min at 37°C place the dish at room temperature for an additional 20 min (up to 40 min).
6. A mixture of 95% alcohol, glacial acetic acid, and 40% formaldehyde mixed in the ratio 6:2:1 is used as fixative. Immerse

68

the coverglass for 6 sec in a dilute solution of fixative (0.1 ml of fixative in 100 ml of 0.17% saline). Place in full strength fixative for 2 to 5 min.

7. Wash 2× in distilled H_2O and allow to air dry at room temperature. The fixed cells now form a flattened tightly adhered monolayer on the glass and can be stored indefinitely.

8. Stain with aceto-orcein or by the Feulgen method (or other chromosome stain, see Chapter 7). The aceto-orcein method is described here. The stain, prepared as described in Chapter 7, should be refiltered daily. Place a drop of aceto-orcein on a microscope slide, place the coverglass cell side down upon it, and apply filter paper to the edges to absorb excess stain. Seal the edges with Kronig cement (Fisher Scientific Co.) applied with a heated spatula. The resulting preparation is now ready for examination, although storage overnight improves it by intensifying the stain. It can be stored in the refrigerator for several months or can be made permanent by placing the slide, coverglass down, on top of a block of dry ice. When the slide is completely chilled, scrape off the cement with a razor blade and then insert the blade gently between the slide and the coverglass to separate them. Place the coverglass in the following series of baths in rapid succession for 3 sec each: absolute alcohol, repeat absolute alcohol, a 1:1 mixture of absolute alcohol and xylol, xylol. Place in xylol for about 5 min. Mount with Permount (Fisher Scientific Co.).

Reference: Tjio, J. H., and Puck, T. T.: Genetics of Somatic Mammalian Cells. II. Chromosomal Constitution of Cells in Tissue Culture. J. Exp. Med., *108*:259, 1958.

6

Analysis of Chromosomes Directly from Tissues

For direct cytogenetic examination of tissues, the cells to be studied must be dividing rapidly at the time of sampling. In some cases, cell dispersion creates a problem if the tissue is dense. A short-term cell incubation may be employed (1) to increase the number of dividing cells and (2) to allow time for an *in vitro* colchicine effect. However, any incubation step inserted between obtaining the tissue and examining the chromosomes raises the criticism that artifact may be introduced.

References: Bartalos, M., and Baramki, T. A.: Tumors and Chromosomes. In: *Medical Cytogenetics.* Baltimore, The Williams & Wilkins Co., 1967, Ch. 22.

Tjio, J. H., and Wang, J.: Direct Chromosome Preparations of Bone Marrow Cells. In: *Human Chromosome Methodology,* J. J. Yunis, ed. New York, Academic Press, 1965.

PREPARATION OF CHROMOSOMES FROM BONE MARROW

The examination of bone marrow is important in many cases of human leukemia when the chromosome constitution must be determined on the marrow stem cells rather than on the peripheral blood.

Direct Examination

1. Aspirate sternal or iliac crest marrow (1 to 2 ml) under local anesthesia.
2. Suspend in 20 ml Earle's solution (balanced salt solution).
3. Dilute 1 part of suspension to 4 parts of hypotonic solution (0.44% Na citrate).
4. Allow to stand 15 min at room temperature.
5. Centrifuge 600 to 800 rpm for about 8 min.
6. Pour off supernatant.
7. Add 50% acetic acid down side of tube to fix cell button for 30 min.
8. Discard fixative. Add 1 ml 2% orcein in 65% acetic acid and allow to stand for at least 10 min. Suspend cells with Pasteur pipette and prepare squash preparation. (See Chapter 7.)

or

Air-dry slides and stain as needed. (See Chapter 7.)

Reference: Sandberg, A. A., Ishihara, T., Kikuchi, Y., and Crosswhite, L. H.: Chromosomal Differences among the Acute Leukemias. Ann. N. Y. Acad. Sci., *113*:663, 1964.

Direct Examination with in vitro Colchicine—Method of Tjio and Wang

SOLUTIONS FOR PRELIMINARY TREATMENT:

Colchicine solution: 0.85% NaCl solution containing 6.6×10^{-3} M phosphate, pH 7, to which is added either colchicine or diacetylmethyl colchicine (Colcemid, Ciba) 0.3 μgm/ml.

Hypotonic solution: Sodium citrate, 1% in distilled H_2O.

STAINING SOLUTIONS:

2% orcein stain: Prepare by dissolving 2 gm orcein (G. T. Gurr's natural or synthetic orcein) in 45 ml hot glacial acetic acid; cool; add 55 ml distilled H_2O; filter.

Orcein-HCl: Orcein stain 9 vol to 1 N HCl 1 vol.

SQUASH PREPARATIONS:

1. Aspirate about 0.5 ml sternal, iliac crest, or tibial marrow and drop immediately into 2 to 3 ml of colchicine solution. Transfer the marrow pieces to a second change of colchicine solution, taking care to free them of blood clots as much as possible. Leave for 1 to 2 hr at 20 to 30°C.

2. Transfer to 2 to 3 ml of hypotonic solution and leave for 20 min at 20 to 30°C.

3. Transfer the material to a watch glass containing a few drops of orcein-HCl and heat it over a small flame to effect rapid fixing, staining, and softening. The solution must not boil.

4. Transfer a piece of marrow to a slide, add a drop of orcein stain, and place a 22-mm square cover slip in position. Tap the cover slip gently and repeatedly with the point of a blunt pencil or needle. Remove excess fluid from the edges of the cover slip with filter paper and express the remainder by pressure applied to the cover slip through blotting paper. Avoid sidewise movements of the cover slip.

5. Seal with Kronig cement (Fisher Scientific Co.). If stored in a cold place, these slides keep about 3 mos. Make permanent by the dry ice method (see Chapter 7, the section on squash preparations).

AIR-DRIED PREPARATIONS:

1. Follow the procedure for squashes, except that the second change of colchicine solution may be omitted if the marrow pieces are very small.

2. Centrifuge at room temperature at 400 rpm for 4 to 5 min. Remove supernatant and add 2 to 3 ml of hypotonic solution. Shake to loosen cells and leave for 0.5 hr.

3. Centrifuge and remove the supernatant. Add 5 ml of fixative (alcohol-acetic, 3:1), resuspend by shaking, and leave for 2 to 5 min. Repeat this procedure a second time.

4. Centrifuge and remove all the supernatant; add approximately 0.2 to 0.5 ml (depending on the volume of the cell button) of fresh fixative and agitate to resuspend the cells.

5. With a pipette put 1 to 2 mm droplets of the cell suspension on chemically clean slides. Blow on each droplet gently as soon as the cells are attached to the glass surface to assist in spreading and drying. Leave to dry thoroughly.

6. Immerse slides 10 to 20 min in a Coplin jar containing orcein stain.

7. Dehydrate in grades of alcohol. Clear in 2 changes of absolute alcohol-xylene (1:1) to avoid excessive extraction of stain, 2 changes of xylene, and mount in a synthetic resin (Permount used). Staining can also be performed with the Feulgen reaction, Wright's or Giemsa stain, or crystal violet (see Chapter 7, on stains).

References: Tjio, J. H., and Wang, J.: Chromosome Preparations of Bone Marrow Cells Without Prior *in Vitro* Culture or *in Vivo* Colchicine Administration. Stain Techn., *37:*17, 1962.

Tjio, J. H., and Wang, J.: Direct Chromosome Preparations of Bone Marrow Cells. In: *Human Chromosome Methodology,* J. J. Yunis, ed. New York, Academic Press, 1965.

Direct Examination with Short-term in vitro Incubation of Cells and Colchicine

(This method was used for human chromosomes prior to development of the short-term peripheral blood method. It is of historical interest.)

1. Disperse marrow specimens in Ringer's solution containing heparin (1:20,000). (Other balanced salt solutions may be used and are now preferred for human cells. See Chapter 12.)

2. Centrifuge and remove supernatant.
3. Resuspend in glucose-saline (0.6% glucose and 0.7% NaCl) plus normal human AB serum. (The more recently defined culture media for mammalian cells plus about 15% serum are now considered more specific for human cells. See Chapter 12.)
4. Incubate at 37°C for 7 hr total incubation, with the addition after 5 hr of colchicine.
5. At the end of the incubation period sufficient 0.37% Na citrate is added to give a fourfold dilution of the medium.
6. After 10 min the cells are fixed and processed for chromosomes.

Reference: Ford, C. E., Jacobs, P. A., and Lajtha, L. G.: Human Somatic Chromosomes. Nature, *181:*1565, 1958.

MODIFICATION

A more recent adaptation of the method to examine chromosomes from bone marrow following short-term *in vitro* incubation (about 7 days) and colchicine has been reported in detail.

Reference: Farnes, P., Barker, B. E., and Fanger, H.: A Technique for Chromosome Study of Human Bone Marrow Fibroblast-like Cells. Exp. Cell Res., *29:*86, 1963.

CULTURE OF BONE MARROW FIBROBLAST-LIKE CELLS

Reference: Punnett, H. H., Kistenmacher, M. L., and Niederer, B. S.: Metachromasia in Fibroblasts. Lancet, *i:*1433, 1968.

PREPARATION OF CHROMOSOMES FROM SOLID TUMORS

A variety of procedures is possible, depending upon the type of tissue to be studied. Several are listed, in order of increasing complexity.

Hypotonic Treatment—Squash Procedure

Place tissue, as fresh as possible, in distilled H_2O, transfer to 45% acetic acid, and make squash preparations. (See Chapter 7.)

Reference: Stolte, L. A. M., v. Kessel, H. I. A. M., Seelen, J. C., and Tijdink, G. A. J.: Chromosomes in Hydatidiform Moles. Lancet, *ii:*1144, 1960.

With in vitro Colchicine

1. Place fresh tissue fragment in culture medium (TC 199) containing 1 μgm/ml colchicine.
2. Mince finely.
3. Incubate 2 hr at 37°C.
4. Replace medium with 0.95% Na citrate solution for 25 min.
5. Fix cells in 1 part glacial acetic acid to 3 parts absolute methanol.
6. Resuspend in 45% acetic acid.
7. Air-dry and stain. (See Chapter 7.)

Reference: Martineau, M.: A Similar Marker Chromosome in Testicular Tumors. Lancet, *i:*839, 1966.

or

1. Tease small fragment of tumor in culture medium (TC 199) containing 1 part per million of Colcemid.
2. Incubate 1 hr at 37°C.
3. Centrifuge. Resuspend in 0.9% Na citrate for 10 min at 37°C.
4. Fix cells by resuspending in 1 part acetic acid to 3 parts ethyl alcohol or by adding acetic acid to the citrate suspension to a final concentration of 5% and then adding ethyl alcohol to a final concentration of 30%.
5. The fixed suspension can be variously handled to produce spreading of chromosomes, either by squashing from 50% acetic acid or by dropping acetic-alcohol suspension on a wet slide. Stain. (Also see Chapter 7.)

Reference: Cox, D., Yuncken, C., and Spriggs, A. I.: Minute Chromatin Bodies in Malignant Tumors of Childhood. Lancet, *ii:*55, 1965.

With in vitro Trypsinization and Colchicine

1. Cut a piece of fresh tumor into small pieces and place, within ½ hr after removal, in a small beaker and cover with Hank's balanced salt solution adjusted to pH 5,* with added Colcemid (4 × 10⁻⁷ M/L) and trypsin (0.25%—Difco 1/250). For many tumors, trypsin is not necessary.
2. Stir tissue gently with magnetic mixer.
3. Decant resulting suspension, consisting largely of single tumor cells, at intervals of 15 to 30 min. Each time add more Hank's

---

* pH 5 was chosen because of the report that this pH produced optimal conversion of rat-liver tissue to a suspension of single liver cells with minimal cell damage.

solution and repeat until sufficient cells have been obtained or the tissue has been fully utilized.

4. Place suspended cells at room temperature in hypotonic Hank's solution containing 160 mg NaCl/100 ml for 30 min (including centrifugation).

5. Process for chromosomes, using standard procedures for cells in suspension. (See Chapter 4, Preparation of Chromosomes from Peripheral Blood, and Chapter 7.)

References: Kotler, S., and Lubs, H. A.: Comparison of Direct and Short Term Tissue Culture Techniques in Determining Solid Tumor Karyotypes. Cancer Res., *27:*1861, 1967.

Lubs, H. A., and Clark, R.: The Chromosome Complement of Human Solid Tumors. New Eng. J. Med., *268:*907, 1963.

With in vitro Trypsinization, Short-term Incubation, and Colchicine

1. Obtain tumor tissue under sterile conditions. Collect in cold Hank's solution, pH 7.4. (See Chapter 12, balanced salt solutions.)

2. Mince to 2 to 3 mm fragments and wash with fresh Hank's solution.

3. Suspend tissue in 0.2% trypsin (Difco 1:250) in phosphate-buffered saline at pH 7.8 and transfer to small Erlenmeyer flask containing a magnetic stirring bar. Sterile technique should be used throughout.

4. Trypsinization is performed at room temperature, and the freshly dispersed cells are decanted every 20 min. Total time varies between 2 and 4 hr, depending on the friability of the tumor.

5. Centrifuge the collected cell suspensions at 1500 rpm for 10 min and combine the sedimented cells in tissue culture medium (199) containing 15% fetal calf serum and penicillin and streptomycin (see Chapter 12).

6. One ml of cell suspension containing approximately 50,000 cells is introduced into each of 15 sterile Leighton tubes with slides (Bellco Biological Glassware and Equipment, Vineland, N. J.).

7. Incubate at 37°C and replace medium after 24 hr and at intervals of 1 or 2 days thereafter, depending on the rapidity of pH change.

8. Prepare samples for cytogenetic study on the 3rd day, and daily thereafter, by adding 1 μgm/ml final concentration of colchicine, and incubating for 5 hr at 37°C.

9. Add an equal volume of prewarmed (37°C) distilled H_2O and incubate at 37°C for 15 min.

10. Fix Leighton slides for 24 hr, using 1 part glacial acetic acid to 3 parts absolute methanol fixative.

11. Hydrolyze with 1 N HCl at 58°C for 7 min and stain using Unna Polychrome Blue (E. F. Mahady Co., Boston). Mount in Permount (Fisher Scientific Co.).

or

Stain with other chromosome stains. (See Chapter 7.)

Reference: Socolow, E. L., Engel, E., Mantooth, L., and Stanbury, J. B.: Chromosomes of Human Thyroid Tumors. Cytogenetics, *3:*394, 1964.

PREPARATION OF CHROMOSOMES FROM TUMOR EFFUSIONS

1. Collect a few ml of pleural or peritoneal fluid in about 10 ml of water containing a small amount of heparin and colchicine to a final concentration of about 0.003%.

2. After ½ hr, spin cells gently, fix, and process as for fixed cells in suspension. (See Chapter 4, Preparation of Chromosomes from Peripheral Blood, and Chapter 7.)

Reference: Jacob, G. F.: Diagnosis of Malignancy by Chromosome Counts. Lancet, *ii:*724, 1961.

MODIFICATION

1. Obtain pleural or peritoneal fluid and concentrate cells by centrifugation.

2. Place a drop of water and a drop of cell suspension on a slide for 20 to 30 min. Prepare by squash technique. (See Chapter 7.)

or

Treat cells in a hypotonic salt solution (1.12% Na citrate), and proceed as for cells in suspension. (See Chapter 4, Preparation of Chromosomes from Peripheral Blood, and Chapter 7.)

Reference: Sharma, A. K., and Sharma, A.: Study of Cancer Chromosomes. In: *Chromosome Techniques.* Washington, Butterworths, 1965, Ch. 11.

SHORT-TERM CULTURES FROM SMALL TISSUE BIOPSIES
(COVERGLASS METHOD)

This technique involves chromosome analysis as soon as the first fibroblast-like cells grow from a primary explant. (Also see Chapter 14 on the handling of primary explants.) If the purpose of the culture is chromosome analysis, considerable time may be saved if the cells are examined before they reach the stage of serial culture.

1. Use sterile technique. Rinse tissue fragments in balanced salt solution and place in tissue culture medium in a Petri dish. (See Chapter 12 for discussion of tissue culture medium and techniques.)
2. Mince tissue to fine cubes of 1 mm or less.
3. Draw 6 to 8 such fragments into a curved Pasteur pipette along with a little medium.
4. Spread them out over the surface of an 11×35 mm coverglass on the floor of a Leighton tube. (See drawing.) Draw off the excess medium.
5. Place a second coverglass over the tissue fragments in such a way that its edge overlaps that of the lower coverglass by a few mm. The two coverslips will adhere together.

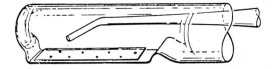 

6. Place each Leighton tube with the explants, flat surface up, on a slightly sloped rack and pipette 2 ml of culture medium down the round surface. (This procedure avoids "floating" the coverglasses apart.)
7. Seal 5 or 6 such tubes with rubber stoppers (if necessary, gas to proper pH with CO_2), and place back on their flat surface in the rack. Incubate at 37 to 37.5°C.
8. Supply the culture with fresh medium only when indicated by a marked pH change of the culture medium. Change is usually indicated on the 5th day for normal tissues and on the 3rd or 4th day for tumor tissues.
9. When a large number of cells are seen to be dividing, add 0.2 ml of 0.01% stock colchicine to each tube.

10. After further 6 hr incubation remove the coverglasses by sliding them out intact from the Leighton tube with the aid of a bent probe.
11. Place coverglasses in a Petri dish containing 5 ml of culture medium at 37°C, separate them, and place cell side up in the dish.
12. Add 15 ml of distilled H_2O drop by drop to the medium in the Petri dish. Leave 10 min.
13. Transfer to fresh Carnoy's or other fixative, culture side up. (See Chapter 7 on fixatives.) Fix 10 min.
14. Air-dry coverglasses on filter paper. Remove tissue fragments while they are still slightly moist.
15. Stain with chromosome stain. (See Chapter 7 on stains.)
16. From explant cultures of normal human skin biopsies, an adequate number of mitotic cells may be obtained for chromosome analysis within 12 days.

Reference: Basrur, P. K., Basrur, V. R., and Gilman, J. P. W.: A Simple Method for Short Term Cultures from Small Biopsies. Exp. Cell Res., *30:*229, 1963.

MODIFICATIONS

1. A standard glass microscope slide may be cut to fit the well in the Leighton tube. Tissue fragments are placed in the tube directly, and the slide (precleaned in 95% alcohol and sterilized separately) is placed on top of the fragments. After culture the fibroblast-like cells growing on the glass slide are processed as described above. Mount the small slide cell side up on a standard size microscope slide.
2. Very small tissue fragments may be put in a plastic tissue culture Petri dish and a sterile coverglass placed over them. The coverglass is anchored as follows: Apply a red hot forceps, flamed in a Bunsen burner, *briefly* to the Petri dish bottom at the edge of the coverglass. Repeat to all 4 sides of the coverglass. The heat raises humps of plastic that keep the coverglass in place.

7

The Handling of Chromosomes

Prior to 1956, the normal total chromosome number for human beings was considered to be 48, instead of the correct 46. Good visualization of individual chromosomes and an awakened interest in human chromosomes were in large measure responsible for correction of the counting error. For a long time colchicine had been used to arrest cells in metaphase and to increase the number of mitoses under examination. Better visualization of individual chromosomes came from the introduction of hypotonic treatment to spread them prior to fixation.

In this chapter the details of preparing the chromosomes for examination will be considered. An effort is made to assimilate the protocols from various laboratories and to discuss the range of difference. It becomes obvious that many procedural modifications are allowable for each laboratory. However, excellent results seem always correlated with meticulous attention to details.

References: Darlington, C. D., and La Cour, L. F.: *The Handling of Chromosomes*. London, George Allen & Unwin Ltd., 1960.

Sharma, A. K., and Sharma, A.: *Chromosome Techniques*. Washington, Butterworths, 1965.

METAPHASE ARRESTING AGENTS

The use of metaphase arresting or "harvesting" agents is variable from one human chromosome protocol to another in regard to (1) type of agent, (2) concentration, and (3) length of application. Higher concentrations and longer application times in general will increase the contraction of metaphase chromosomes. In the presence of colchicine, after a period of time, normal mitoses may be completed or *endoreduplication* may occur. Colchicine metaphase arrest is also reversible if the colchicine is removed.

A range of procedures for human chromosomes is listed in this section. Specific recommendations are given under the specific complete protocols in other chapters.

Type of Metaphase Arresting Agent

(All types prevent spindle formation or alter it.)

Colchicine—has been in use the longest.

Deacetylmethylcolchicine (Colcemid-Ciba)—especially for *in vitro* use, less toxic than colchicine.

Vinblastine sulfate (Velban)

Concentration and Time (in vitro)

Colchicine		Colcemid		Velban	
0.5 μgm/ml	1 to 6 hr	2 μgm/ml	1 to 6 hr	0.1 μgm/ml	1 to 3 hr
0.1 μgm/ml	1 to 6 hr	0.06 μgm/ml	4 hr	0.01 μgm/ml up to 6 hr	
0.05 μgm/ml	1 to 6 hr or longer	0.03 μgm/ml	2 hr		

References: Eigsti, O. J., and Dustin, P.: *Colchicine.* Ames, Iowa, The Iowa State College Press, 1957.

Krishan, A.: Time-Lapse and Ultrastructure Studies on the Reversal of Mitotic Arrest Induced by Vinblastine Sulfate in Earle's L Cells. J. Nat. Cancer Inst., *41:*581, 1968.

HYPOTONIC TREATMENTS FOR CHROMOSOME SPREAD

The hypotonic treatments vary from one human chromosome protocol to another in regard to (1) hypotonicity and composition, (2) length of application, and (3) temperature. The spreading response seems variable even when all three are presumably constant. The ease of spreading chromosomes differs from one mammalian species to another. Furthermore, fixation and drying procedures, as well as the process of metaphase arrest while the cells are still in culture, also influence chromosome spread. Lastly, chromosomes obtained from poorly growing cells are usually difficult to spread well.

A range of procedures for human chromosomes is listed here. Specific recommendations are given under the specific complete protocols in other chapters.

Hypotonic Solutions

Distilled H_2O

3, 4, or 5 parts distilled H_2O to 1 part isotonic solution or medium

0.17% saline in distilled H_2O

4 parts 0.17% saline to 1 part isotonic solution or medium

1% sodium citrate or less

Length of Application

3 to 40 min

Temperature

Room: 23°C

37°C

Reference: Hungerford, D. A., and DiBerardino, M.: Cytological Effects of Prefixation Treatment. J. Biophys. Biochem. Cytol., *4:*391, 1958.

FIXATIVES FOR THE EXAMINATION OF CHROMOSOMES

Acetic Alcohol

(perhaps the most universal fixative for chromosomes)

1. 3 parts absolute methanol* to 1 part glacial acetic acid.
2. Glacial acetic acid is the heavier component. To insure a well-mixed fixative, it is easier to add the glacial acetic acid second.
3. Prepare fresh and mix well.
4. When the fixative is applied to a button of cells, add slowly down the side of the tube, without disturbing the button. By this method the penetration of the fixative is slower.

Acetic Acid

45% Acetic acid in H_2O—to increase nuclear and chromosome spread.

50% Acetic acid in H_2O—used in some protocols instead of 45% acetic acid to increase chromosome spread.

Carnoy's Fixative (by volume)

100 absolute alcohol
16 glacial acetic acid
50 chloroform

Tjio Fixative †

95% alcohol 6 parts
glacial acetic acid 2 parts
40% formaldehyde 1 part

Time of Fixation

The time of fixation varies with the type of specimen. A button of cells is usually fixed for not less than 30 min, preferably with slow penetration. However, in some protocols fixative is added slowly to the cells while they are still suspended in prefixation solutions (such as hypotonic solution). Storage of human cells in fixative for prolonged periods of

---

* Absolute ethanol is used in some protocols.

† Tjio recommends diluted fixative step first for cells on coverglass. (See Chapter 5, Preparation of Chromosomes from Human Cells Growing on Coverglasses.)

time is in general not recommended, particularly if the fixative is acetic acid (which keeps increasing the chromosome spread). Storage of a button of cells in 1:3 fixative, in the refrigerator, at least overnight, has been recommended.

Cells on coverglasses or slides are usually fixed for not less than 5 min or more than 24 hr. Prolonged fixation tends to result in loss of cells.

AIR-DRYING TECHNIQUE FOR CHROMOSOME SPREAD

Air drying is probably the simplest procedure for spreading chromosomes on a microscope slide from a suspension of cells in fixative. Many slides can be prepared quickly by this procedure. It is recommended for students who are preparing their own slides. It is also recommended for autoradiography of chromosomes, to facilitate scanning of slides since the dried cell suspension can be easily followed down the slide. Furthermore, if it is desirable to sample a large proportion of all of a specimen, then multiple (up to three) drops can be placed side by side on a slide. This method has the disadvantage that the chromosomes in some human preparations and in many preparations from other mammals are not as well spread as with the flame-drying or squash techniques also described in this chapter.

Procedure

1. Slides should be precleaned by soaking in 95% alcohol. Wipe dry with lint free cloth.
2. Cell suspension should be moderately heavy. Cells may be in either 1 part glacial acetic acid:3 parts absolute methanol or in 45% acetic acid. At least one change of 1:3 fixative is recommended prior to air drying from this fixative. If 45% acetic acid is used to increase the chromosome spread, it should follow 1:3 fixation. Once the cells are in 45% acetic acid it is difficult to obtain a button again after centrifugation. Therefore it is important to avoid an excess volume of 45% acetic acid fixative.
3. A Pasteur pipette with a 1-ml rubber bulb is used. The cell suspension should fill the tapered end without interruptions. A drop is allowed to fall from a height of several inches down onto the slide near the end with the label. The slide is held at a 60° to 90° angle and the drop of cell suspension is allowed to run down the slide quickly. The excess liquid (if any) may be

removed from the end of the slide by touching it quickly to soft blotting paper. The slides are allowed to dry completely in the slanted position before staining or storage.

CHROMOSOME SPREAD BY FLAME DRYING

1. Place slides precleaned in alcohol in distilled H_2O at refrigerator temperature (overnight).
2. After fixation of cells in 1:3 acetic alcohol, make a suspension in fresh fixative. (A cell button of .05 ml volume should be suspended in about 0.5 ml or less of fixative. Keep cell suspension heavy at the beginning, since more fixative can be added if cells are too thick on the slide.)
3. Assemble slides, cell suspension, and Bunsen burner.
4. Remove one slide from the cold water; drain off excess water on towel or blotter briefly. Hold slide level in one hand.
5. With the other hand mix the cell suspension; with Pasteur pipette and 1-ml rubber bulb, draw up a small amount, hold pipette tip about 8 inches above slide, and drop 1 drop, then another drop about 1 inch apart on slide. The drops will splash, and fixative and water will mix rapidly.
6. Drain off excess liquid briefly, and pass slide through the flame to dry rapidly. Do not get the slide hot. A good way to avoid overheating is to hold your entire hand over the flame, at a distance which is not uncomfortably hot. Wave the slide back and forth at this distance from the flame.

PREPARATION OF CHROMOSOMES BY THE SQUASH TECHNIQUE— BOTH QUICK AND PERMANENT MOUNTING METHODS

In addition to the air- and flame-drying techniques that have already been described, the squash technique remains one of the most satisfactory methods to spread and flatten metaphase chromosomes for temporary or permanent mounting.

1. Place a small drop of heavy cell suspension in fixative (1:3 glacial acetic acid:absolute methanol or 45% acetic acid) on a microscope slide precleaned in alcohol.
2. Place a small drop of aceto-orcein (2% in 45% acetic acid) on top of the cells.*

---

* Alternatively the fixed cells may be stained as a button in the bottom of a centrifuge tube.

85

3. Place a square coverglass over the cells and apply steady thumb pressure. A piece of blotting paper may be used on top of the coverglass, but care must be taken to avoid sideward slipping of the coverglass. The correct amount of pressure can be judged by examining the cells under the microscope after a preliminary squash.

4. A temporary preparation is made by ringing the coverglass with Kronig cement (Fisher Scientific Co.) applied with a heated spatula.

5. The orcein stain darkens overnight. This preparation may be stored at refrigerator temperature for several months.

6. A permanent preparation is made by placing the slide, coverglass down, on top of a piece of dry ice. The frozen cement can then be scraped off with a sharp instrument, and the coverglass can be lifted. Nearly all the squashed material sticks to the slide when the coverglass is removed.

7. The slide and coverglass are placed immediately in absolute alcohol for about 5 min, absolute alcohol-xylene and mounted from xylene.

References: Conger, A. D., and Fairchild, L. M.: A Quick-Freeze Method for Making Smear Slides Permanent. Stain Techn., *28:*281, 1953.

Sachs, L.: Simple Methods for Mammalian Chromosomes. Stain Techn., *28:*169, 1953.

CHROMOSOME STAINS

Many chromosome stains are available, with varying degrees of specificity for DNA. The Feulgen reaction is usually considered the most specific, since in this procedure DNA after acid hydrolysis liberates aldehyde groups that react with leuco-basic fuchsin, resulting in a red-violet coloration of the chromosomes. However, chromosomes are stained very satisfactorily by less specific stains, and the choice of which one to use is often based on such considerations as (1) how stable is the stock solution to prolonged storage, (2) how reproducible is the staining procedure, and (3) what are the particular preferences or habits in a given laboratory. Leuco-basic fuchsin can produce a beautiful result, but the procedure is notoriously difficult to reproduce and the stock is unstable to light. Giemsa stain is stable in stock solution and is highly reproducible, but may result in some loss of detail such as secondary constriction or coiled chromatid substructure. Procedures for some of the possible stains are described here.

References: Darlington, C. D., and La Cour, L. F.: *The Handling of Chromosomes.* London, George Allen & Unwin Ltd., 1960.

Gurr, E.: *Staining: Practical and Theoretical.* Baltimore, The Williams & Wilkins Co., 1962.

Sharma, A. K., and Sharma, A.: *Chromosome Techniques.* Washington, Butterworths, 1965.

Aceto-orcein Stain

1. Dissolve 2 gm orcein (natural or synthetic) in 45 ml hot glacial acetic acid by boiling gently (under hood) with stirring for 3 to 5 min. Cool and add 55 ml distilled H_2O to make 2% solution in 45% acetic acid. Filter. Refilter daily, as needed.
2. Staining time varies with type of orcein and desired intensity. 10 to 20 min is usual for slides and coverglasses.
3. This stain is useful for temporary and permanent chromosome squash preparations. (See section in this chapter on squash preparations.)

Carbolfuchsin Stain

Reference: Carr, D. H., and Walker, J. E.: Carbol Fuchsin as a Stain for Human Chromosomes. Stain Techn., *36:*233, 1961.

Eosin-Stevenel's Blue Stain

REAGENTS

1. 1 gm eosin; 500 ml distilled H_2O.
2. Stevenel's blue. Mix 1 gm methylene blue in 75 ml distilled H_2O with 1.5 gm potassium permanganate in 75 ml distilled H_2O. Keep in H_2O bath at 100°C for 30 min, and filter after 24 hr.

STAIN

1. Immerse slide in alcohol for 10 sec, and then air dry.
2. Immerse slide in eosin for 30 sec, in Stevenel's blue for 30 sec, in eosin for 10 sec, and Stevenel's blue for 10 sec. Rinse with running tap H_2O between each immersion.
3. Air dry and mount.

Reference: Grove, S.: A Simple Chromosome Stain. Lancet, *ii:*1146, 1967.

Feulgen Reaction

LEUCO-BASIC FUCHSIN

1. Dissolve 1 gm basic fuchsin by pouring over it 200 ml boiling distilled H_2O.
2. Shake well and cool to 50°C.
3. Filter: Add 30 ml N HCl to filtrate.
4. Add 3 gm $K_2S_2O_5$.
5. Allow solution to bleach for 24 hr in a tightly stoppered bottle in the dark; add 0.5 gm decolorizing carbon (Norit).
6. Shake well for about a minute and filter rapidly through coarse filter paper.
7. Store in tightly-stoppered bottle in the dark, at 4°C.

SO$_2$ WATER

1. 5 ml N HCl.
2. 5 ml $K_2S_2O_5$ 10%.
3. 100 ml distilled H_2O.

PROCEDURE

1. Rinse slides in distilled H_2O.*
2. Hydrolyze in N HCl at 60°C for about 8 min. (This time may vary slightly depending on the fixative used.)
3. Stain in leuco-basic fuchsin for 1 to 2 hr.
4. Fresh SO$_2$ water in covered jars, 3 changes, each of 10 min (or less).
5. Distilled H_2O rinse. Air dry. Mount from xylene or run through alcohols to xylene and mount.

Fluorescent Staining (Coverslip Method is described.)

1. Subculture cells for about 48 hr on coverslips to form a monolayer.
2. Add colchicine 10^{-6} for 16 hr.
3. Introduce ¼ strength balanced salt solution for 10 min.
4. Fix with 1 part glacial acetic acid to 3 parts absolute alcohol for 15 min.
5. Air dry.
6. Stain with acridine orange (Gurr, London, England) 1:1000 in absolute alcohol for 10 min.

---

* The Feulgen reaction may also be applied to a button of cells.

88

7. Wash in phosphate buffer at pH 7.6.
8. Mount in liquid paraffin or ring with "Edging Lacker" (Zeiss).
9. Examine on standard fluorescent microscope, using a high-discharge mercury vapor lamp (Zeiss HBO 200). Two exciter filters (Zeiss BG 12) give a range of 3800 Å and 2 barrier filters (Zeiss OG 5) give good differentiation. The chromosomes appear bright yellow, while the cytoplasm shows a red fluorescence.

Reference: Salkinder, M., and Gear, J. H. S.: Fluorescent Staining of Chromosomes. Lancet, *i:*107, 1962.

Giemsa Stain

Dilute from commercial stock solution by adding 2 ml of stock stain, 2 ml phosphate buffer pH 6.4, to make 50 ml in distilled H_2O.

PHOSPHATE BUFFER pH 6.4

Prepare as described for tetrachrome stain or else prepare as follows:

1. Solution A: KH_2PO_4 11.336 gm/100 ml H_2O *or*
                              56.68   gm/500 ml H_2O.
2. Solution B: Na_2HPO_4 8.662 gm/100 ml H_2O *or*
                              43.31   gm/500 ml H_2O.
3. 5.0 ml solution A.
   5.0 ml solution B.
   Make to 1000 ml with distilled H_2O.
4. Adjust pH to 6.4 with 0.1 N HCl.

GIEMSA CONCENTRATE (MAY BE PREPARED AS FOLLOWS:)

Giemsa stain, powder   3.8 gm (1.0 gm)
Glycerine   250.0 ml (66.0 ml)
Maintain at 55 to 60°C for 1½ to 2 hr, and then add:
Methyl alcohol   250.0 ml (66.0 ml)

STAINING

Staining time for slides varies from about 2 min to 10 min. Rinse 2X in distilled H_2O or rinse gently in running distilled H_2O. Wipe off back of slide and air dry.

Safranin Stain

Reference: Smith, K. D., Steinberger, A., and Perloff, W. H.: Safranin as a Stain for Chromosomes. Lancet, *i:*1379, 1963.

Tetrachrome Stain (MacNeal)

1. Mix the following dry ingredients:
   1.00 gm methylene blue chloride
   0.60 gm azure A
   0.20 gm methylene violet
   1.00 gm eosin Y

   or

   Tetrachrome stain-MacNeal is available commercially.
2. Dissolve 0.150 gm of dry ingredients in 100 ml of methyl alcohol, neutral, acetone free, by heating to 50°C with shaking.
3. Leave 1 day at 37°C with occasional shaking.
4. Filter off any precipitate and store in dark, airtight bottle.
5. Pipette 1.0 ml stain into test tube and add 2.0 ml phosphate buffer pH 6.4 (2.53 gm Na_2HPO_4, 6.65 gm KH_2PO_4 to 1 liter distilled H_2O or prepare phosphate buffer pH 6.4 as described for Giemsa stain).
6. Shake gently to mix and pour onto slide. Stain 2 or more min.
7. Wash stain from slide by pouring distilled H_2O over slide.
8. Blot dry gently or air dry.

Wright Stain

Reference: Tubiash, H. S.: A Rapid, Permanent Wright's Staining Method for Chromosomes and Cell Nuclei. Amer. J. Vet. Res., *22:*807, 1961.

MOUNTING

Xylene is the most universal solvent for mounting materials. (However, see special instructions accompanying one you have not used before.) Slide or coverslips may be air dried thoroughly and mounted following a brief immersion in xylene. Slides or coverslips wet with stains in water or alcohol bases must be moved through alcohols to xylene: 70%, 95% absolute, xylene. Some protocols recommend starting from water into alcohol more dilute than 70%. Absolute alcohol-xylene (1:1) may also be inserted between absolute alcohol and xylene. Permount and Histoclad are commonly used synthetic mounting media.

In many instances mounting is unnecessary and, in fact, contraindicated. Unmounted slides are easily destained and restained as needed. Unmounted slides with immersion oil on them can be cleaned easily by soaking in xylene and blotting dry between standard blotting paper.

Temporary mounts (which can be made permanent) are described in this chapter in the section on preparation of chromosomes by the squash technique.

TREATMENTS TO BRING OUT SPECIAL DETAILS OF CHROMOSOME STRUCTURE

Demonstration of Spiral Structure

1. Treat 3-day peripheral blood cultures with colchicine 0.06 $\mu g/ml$ (or more) for 1 hr (or longer) at 37°C. Suspend cells by shaking. Centrifuge and discard supernatant.

    or

    Follow culture procedures for monolayer cells (Chapter 5 and Chapter 12). Add colchicine 0.06 $\mu g/ml$ (or more) for 1 hr (or longer). Decant medium and save in centrifuge tube. Spin medium 8 min at 600 to 800 rpm. Pour off supernatant. Add 3 ml PBS (buffered salt solution) to cell monolayer, rotate, pour off into centrifuge tube, add 3 ml PBS, rotate, pour off into same centrifuge tube. Add 1 ml 0.25% trypsin-0.25% Versene and incubate 1-plus min till cells come off. Add to same centrifuge tube. Spin 8 min at 600 to 800 rpm. Pour off supernatant. Remove last amount with Pasteur pipette.
2. Add 2 to 3 ml despiralizing hypotonic mix.* Allow to stand at room temperature for 90 min after gentle suspension of cells with Pasteur pipette.
3. Add 5 to 6 ml cold, fresh fixative (1 part glacial acetic to 3 parts absolute methanol) to hypotonic mix and cells. Mix gently. Fix for 30 min.
4. Centrifuge. Procedure A: Change fixative 3 times. Prepare flame-dried slides. Procedure B: After 3 fixative changes, re-

---

* Despiralizing hypotonic mix:

KCl	7.45 gm/200 ml = 0.5 M	Dilute 5.5 ml of each to 50 ml to obtain .055 M solutions.
NaNO$_3$	8.5 gm/200 ml = 0.5 M	
CH$_3$COONa	8.2 gm/200 ml = 0.5 M	

.055 M KCl	4 parts	Mix.
.055 M NaNO$_3$	2 parts	
.055 M CH$_3$COONa	1 part	

suspend, centrifuge, add small amount of 45% acetic acid. Prepare drip-dried slides.

5. Stain lightly with Giemsa. Wash with running H_2O for a few seconds. Air dry thoroughly.

Reference: Ohnuki, Y.: Demonstration of the Spiral Structure of Human Chromosomes. Nature, *208:*916, 1965.

Enhancement of Secondary Constrictions

The short-term culture of cells in calcium-free medium causes *secondary constrictions* of human chromosomes to appear as nonstaining gaps in specific locations on certain chromosomes.

PROCEDURE FOR MONOLAYER CELLS

1. Wash the cells in 3 changes of calcium-free medium prewarmed to 37°C. Add same medium.
2. Incubate cells with colchicine (0.1 $\mu g/ml$) for 1 to 1½ hr.
3. Follow procedure for chromosomes from human cells in long-term culture (Chapter 5).

Reference: Sasaki, M. S., and Makino, S.: The Demonstration of Secondary Constrictions in Human Chromosomes by Means of a New Technique. Amer. J. Hum. Genet., *15:*24, 1963.

THE ISOLATION OF CHROMOSOMES

Several laboratories have considerable experience with the isolation of metaphase chromosomes. Extension of metaphase chromosomes back to a metabolically active stage has proved to be difficult.

References: Huberman, J. A., and Attardi, G.: Isolation of Metaphase Chromosomes from HeLa Cells. J. Cell Biol., *31:*95, 1966.

Maio, J. J., and Schildkraut, C. L.: A Method for the Isolation of Mammalian Metaphase Chromosomes. In: *Methods in Cell Physiology,* Vol. II, D. M. Prescott, ed. New York, Academic Press, 1966.

Salzman, N. P., Moore, D. E., and Mendelsohn, J.: Isolation and Characterization of Human Metaphase Chromosomes. Proc. Nat. Acad. Sci., U.S.A., *56:*1449, 1966.

Sommers, C. E., Cole, A., and Hsu, T. C.: Isolation of Chromosomes. Exp. Cell Res., Suppl. *9:*220, 1963.

8

Microscopy and the Examination of Chromosomes

This chapter covers photography and examination of human metaphase chromosomes through the bright field light microscope. References to the study of mammalian chromosomes by other types of light microscopy and by electron microscopy are provided at the end of the chapter. Some details of photomicroscopy are included for those who want to start from the beginning or who want to revise procedures. Eye *karyotyping* is explained extensively in order to give it the respectability it deserves. *Autoradiography* is discussed in a separate chapter.

References: Bartalos, M., and Baramki, T. A.: *Medical Cytogenetics.* Baltimore, The Williams & Wilkins Co., 1967, Ch. 2.

Christenson, L. P.: Applied Photography in Chromosome Studies. In: *Human Chromosome Methodology,* J. J. Yunis, ed. New York, Academic Press, 1965.

Needham, G. H.: *The Practical Use of the Microscope.* Springfield, Ill., Charles C Thomas, 1958.

Patau, K.: Identification of Chromosomes. In: *Human Chromosome Methodology,* J. J. Yunis, ed. New York, Academic Press, 1965.

94

EYE KARYOTYPES

One of the shortcuts in chromosome analysis is the eye karyotype. This method is in much more common practice than cytogenetics texts would indicate. A metaphase spread is examined with oil immersion magnification (which is usually about 1000X), and the chromosome analysis is completed by rough drawing and without photography. This method has the advantage that chromosomes are carefully examined on the slide and distortions of photography are not introduced. The microscope focus can be adjusted for each chromosome or parts of chromosomes. Disadvantages are that a permanent photographic record is not made and the relative lengths of various chromosomes may be more difficult to compare when the chromosomes in question cannot be placed side by side. Depending on the purpose of the chromosome analysis, the eye karyotype method may be used exclusively or may be combined with photography. The outline presented here is recommended as a routine procedure for a clinical cytogenetics laboratory. However, with slight modifications the method is also suitable for any more complicated clinical or research cytogenetic problem. There is no question that considerable time may be saved by this method.

1. A knowledge of the classification of normal and abnormal human chromosomes (Chapter 3) is required at the onset. It is a good idea to keep a sample of the normal human male karyotype by the microscope for reference.

2. With low power select a well spread metaphase, and then go directly to oil immersion. Time is saved if the slide is not mounted. The stain can be repeated or modified more easily if the coverglass is not present. Since the examination is with oil, the refractive index of mounting medium is not required. Unmounted slides with oil on them should be cleaned immediately after use by immersion for several minutes in clean xylene. They should be blotted dry carefully.

3. Again review the quality of the metaphase at high magnification. Reject the cell if there is poor spread, too much spread, too much loss of the circular configuration of the spread, overlapping of chromosomes, poor staining of chromosomes, or another metaphase too close. There is some disagreement among cytogeneticists as to what degree of chromosome condensation (or extension) is the best for analysis. The sample metaphase (Fig. 1-2, page 5) illustrates the stage of metaphase usually analyzed.

4. Prepare a quick line drawing of each chromosome. Retain the orientation of the chromosomes and their relative lengths in the drawing as they are on the slide. Be sure to place the slide identification and location of the metaphase at the top of each drawing. It is usually best to make one drawing per page. These pages are stored in folders and labeled according to culture number or name.

5. While the drawing is being made, keep a running count of the chromosome number. Place this number at the bottom of the page or else keep it in mind. It will be checked again as the analysis proceeds.

6. To one side of the line drawing write in a vertical column the numbers of the chromosome pairs that can be identified; otherwise, the group letter. (See Fig. 8-1.) X chromosomes are included in the C group and Y chromosome in the G group. If the Y is distinguishable, it can be noted last.

7. Examine the metaphase carefully. Find the number 1, 2, and 3 chromosomes. Write these numbers next to the chromosome line on the drawing. Also record each chromosome in the column next to the drawing. Then locate the B chromosomes and record

Slide 120-1 5-2-68
Location 27.4/101.3

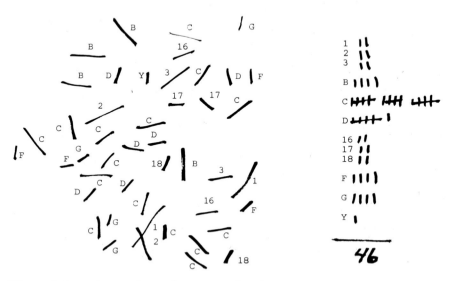

Figure 8-1. *A sample eye karyotype drawing.*

in a similar way. Then identify G chromosomes (and Y), F, and D. Follow with 16, 17, and 18. Finally, identify the C chromosomes. Total the number of chromosomes you have identified. The number should correspond to your original count. Be sure you have written a number or letter after every chromosome line on the drawing. Also be sure you have a line on the drawing to correspond to every chromosome on the slide. These 2 types of errors are commonly made.

8. If there is any problem in the assignment of numbers or letters to the various chromosomes, try to identify the problem. Is a chromosome extra? Which one? Is a chromosome the wrong size? What is wrong about the size? Make a note of what you think is wrong, note that this metaphase may need to be photographed for further analysis, and proceed to another metaphase.

9. In a routine analysis, prepare 20 drawings. If there have been no problems with the chromosome assignments and if the analysis is normal, no further work on this analysis is necessary. Prepare no photographs unless, for some reason, a photograph is

97

needed for the patient's record. Routine photography usually wastes time. Since the locations of good metaphases are recorded and the slides are kept on file, photographs may be taken at a later date, if required.

10. When a consistent abnormality is found in every drawing, say an extra G group chromosome, 20 drawings without photography are sufficient unless the findings are in complete discord with the clinical findings. In such a case, the fault is likely to be mixing of specimens, rather than errors in the eye karyotyping. Therefore, another specimen should be obtained.

11. If: (a) inconsistent abnormalities are found; (b) a consistent abnormality is present in some but not all of the cells; or (c) the abnormality is unusual or not suspected clinically, then each case must be individualized. A few general rules are suggested here.

Inconsistent Abnormalities

In many cases inconsistent abnormalities involve number (whole chromosomes). The cause is usually broken or overlapping cells. Another person experienced in analysis should review the cells and drawings. If indicated, about 50 total counts should be obtained and recorded as in Table 8-1.

Table 8-1. Recording of Total Chromosome Numbers

No. of chromosomes

<45	45	46	47	48	>48	
6	1	43				No. of cells

Eye karyotypes should be prepared of the additional metaphases with abnormal numbers. Review of these drawings *may* suggest the need for karyotyping of photographs. If chromosomes are lost or gained randomly, the cause is *usually* broken or overlapping cells.

Consistent Abnormality in Some but Not All Cells

Mosaicism in the patient is strongly suggested when there is consistent abnormality in some but not all of the cells. At least 50 metaphases should be counted. The additional deviant cells should have drawings. Photography may be indicated. Sampling of multiple tissues may also be indicated.

98

Unusual Abnormality or One Not Suspected Clinically

Additional types of chromosome study are indicated if the abnormality is unusual or not suspected clinically, especially if definitive identification makes a difference to the patient. These analyses include (a) another specimen of the same tissue or multiple tissue sampling, (b) photography and karyotyping, (c) autoradiography, (d) evaluation of other family members.

KARYOTYPES FROM PHOTOGRAPHS

Chapter 3, particularly Table 3-1, describes the arrangement of normal human chromosomes. A sample normal karyotype should also be available for frequent reference (Fig. 1-3, page 6). Samples of abnormal karyotypes (Chapter 2) may be helpful from time to time.

1. At least 2 photographs of each metaphase should be available (3 or 4 if overlapping chromosomes are present). One photograph is cut apart, and the other is kept intact for orientation and reference or for recheck with the original metaphase on the slide. The intact photograph is also useful if a cutout chromosome is misplaced.

2. Cut around individual chromosomes, but if overlapping chromosomes are present, remove the whole configuration together, and separate after the other chromosomes are arranged. The chromosome(s) thereby cut into pieces can be located on the extra photographs and cut from them.

3. The initial arrangement can be made on any flat surface.

4. Methods used to keep the chromosomes in place vary, depending on the purpose of the karyotype. Except for publication, a standard format is usually a good idea. (Fig. 8-2; the X chromosome(s) and Y may be placed in the C and G groups respectively.)

   a. Simply place strips of cellulose tape over the chromosomes. They can still be rearranged to some extent by lifting up the cellulose tape.

   b. Use double-faced cellulose tape or masking tape. If this method is used, karyotypes stored one on top of the other will stick together. Clear plastic may be placed over each karyotype.

A(1-3)   B(4-5)

C(6-12)

D(13-15)   E(16-18)

F(19-20)   G(21-22)

Figure 8-2. *A sample format for preparing a karyotype from a photograph.*

    c. Glue down individual chromosomes. This method is time-consuming; it is necessary, however, if the karyotype is to be published or placed on display as a permanent record. Rub-on letters and numbers or strip symbols made on a machine are recommended (most audiovisual departments can supply both of these kinds of inscriptions).

5. On the back of each karyotype record the roll and frame numbers of 35 mm film or the nomenclature of sheet film. It is also helpful to record the culture number (or patient's name), the slide identification, and the location on the slide. Just the film identification is acceptable as a shortcut. In such a case an exact photography record is important. (See section, this chapter, on photography records.)

6. The importance of returning to the metaphase on the slide during the preparation of the karyotype cannot be overemphasized. Individual chromosomes can be studied in detail on the slide and compared to the photograph. On the slide, the focus can be

individualized for each questionable chromosome; centromeres can be identified with more certainty; blurred chromosomes may be resolved.

7. Interpretation: Some general suggestions are covered under the section in this chapter on eye karyotypes. Chapter 2 on human chromosome variations and abnormalities should also be consulted.

THE PHOTOMICROSCOPE

The assembly of a satisfactory microscope for both chromosome analysis and photography is accomplished best by each individual investigator in consultation with a microscope salesman or representative who knows his equipment. Many cytogenetic laboratories choose an assembly made by either the Leitz or Zeiss companies. Two assemblies will be described in detail to serve as illustrations.

> *References:*   Leitz and Zeiss instruction manuals.
> Needham, G. H.: *The Practical Use of the Microscope.* Springfield, Ill., Charles C Thomas, 1958, Ch. XXVI, Photomicroscopy.

Leitz Photomicroscope Assembly for 4″ by 5″ Film (Fig. 8-3)

EQUIPMENT

Ortholux microscope
Aristophot photomicrographic apparatus (complete for 4″ × 5″ film)*
Graduated mechanical stage
100X Plano objective
40X (Apochromatic) objective
10X, 25X Achromatic objectives
10X eyepieces, high eyepoint (for use with corrective eyeglasses)
10X eyepieces, wide-field
6 (or more) 4″ × 5″ double film holders
Microsix-L photometer
(Phase accessories: condenser, focusing telescope, filters, objectives)

---

* Polaroid or 35 mm may be used.

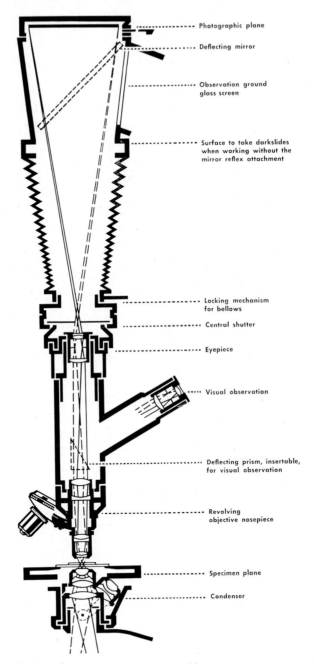

- - - - - - Photographic plane

- - - - - - Deflecting mirror

- - - - - - Observation ground
glass screen

- - - - - - Surface to take darkslides
when working without the
mirror reflex attachment

- - - - - - Locking mechanism
for bellows

- - - - - - Central shutter

- - - - - - Eyepiece

- - - - - - Visual observation

- - - - - - Deflecting prism, insertable,
for visual observation

- - - - - - Revolving
objective nosepiece

- - - - - - Specimen plane

- - - - - - Condenser

Figure 8-3. *Leitz photomicroscope assembly.*

102

PREPARING THE SLIDE

1. Select metaphase, place oil on slide, 100X oil immersion objective.
2. Flip-in lens, in, with oil cap, oil on cap, condenser in position so that oil touches bottom of slide.
3. Close field diaphragm all the way and adjust height of the condenser until the image of the field is sharp.
4. Center condenser with 2 centering knobs by observing edge of field diaphragm.
5. Open field diaphragm until the field is completely illuminated.
6. Close condenser diaphragm about ¾ closed.

LOADING THE FILM (4″ × 5″)

1. Make sure the handles of the film holder covers are shiny side out (indicating that the film is unexposed).
2. In darkroom in total darkness, pull film holder cover part way out.
3. Hold film so that the notches are in the upper right corner.
4. Put film* in the holder so that film is under the guides provided in the holder.
5. Insert cover the rest of the way and close the latch.
6. Do the same on the other side and for multiple film holders.
7. By the light in the microscope room, insert the loaded film holder into the camera. Make sure that the top "snaps," indicating that the film holder is firmly in place.
8. Store extra loaded film holders in a specially assigned dark place. Do not mix with film holders that are unloaded or exposed.

TAKING THE PICTURE

1. Remove right ocular. Insert photometer eyepiece and note reading. The photometer setting is determined by the type of film and the type of filters on the microscope. The seconds reading must be corrected for the bellows height (see instructions with the equipment).
2. Put the light through the camera and set dial to T (time).
3. Push mirror handle away, open shutter by pushing shutter release, and focus image on ground glass plate with fine adjustment of microscope. (Turn off lights and remove filters to do this step. Be sure to replace the filters afterwards.)

---

* 4″ × 5″ sheets, Contrast Process Panchromatic Film, Eastman Kodak, supplied in boxes of 25 sheets.

4. Close shutter and pull mirror handle toward you.
5. Remove lower cover on film holder.
6. Expose appropriate number of seconds (as determined under 1).
7. Replace cover on film holder (with dark side of cover handle out, to indicate that the film is exposed).

GENERAL COMMENTS

The Leitz photomicroscope assembly produces excellent results and chromosome photographs of high quality. $4'' \times 5''$ film enlarges better than 35 mm film. However, for a large amount of routine photography this setup is time-consuming. The image is focused on a ground glass plate that requires a dark environment. The exposure is not automatic.

Zeiss Photomicroscope Assembly for 35 mm Film (Fig. 8-4)

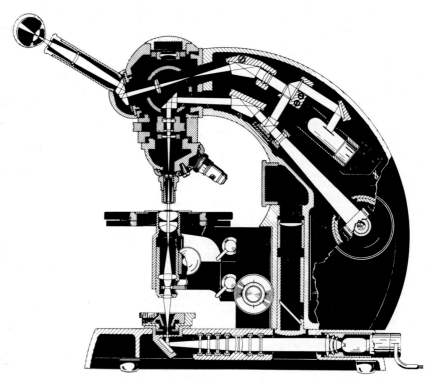

Figure 8-4. *Zeiss photomicroscope assembly. (Courtesy of Carl Zeiss, Inc., New York.)*

104

Equipment

Photomicroscope, with substage condenser and auxiliary lenses
Automatic exposure device
Graduated mechanical stage
Planachromat objectives (2.5X), 6.3X, 16X, 40X, 100X oil*
Eyepieces 10X
2 (or more) film cassettes, 35 mm
(Phase accessories)

Procedure

1. Load 35 mm cassette† (see accompanying instructions and pictures), and place in photomicroscope.
2. Wind advancing spring on camera.
3. Turn light switch on (red light shows on automatic exposure device). Put filters in place if they are to be used. (Filters for various stains are listed on page 108.)
4. Expose 3 frames by pressing the shutter button (B) on the automatic exposure device which keeps the camera shutter open as long as the button is depressed.
5. Set film frame counter on 1.
6. Place slide on the mechanical stage and locate the desired cell under oil immersion (100X). Focus well.
7. A drop of immersion oil may be placed on the condenser "swing-out" front lens under the slide. (This step is optional for routine work.)
8. Note measurement of the interpupillary distance and set the eyepieces for this number.
9. Now set lenses and diaphragms for photography with 100X objective. (See the Zeiss instruction tables.)
   a. Swing-out front lens IN
   b. Auxiliary lens OUT
   c. Condenser diaphragm OPEN (to be adjusted later for contrast)
   d. Front diaphragm OPEN

---

* Planachromat, Achromat, Neofluar, and Planapochromat 100X objectives were compared. The Planachromat objective was found to be the best for the price for chromosome analysis.

† Plus-X Panchromatic film, Eastman Kodak. Buy 100 ft rolls and roll your own. (See section in this chapter on film.)

105

10. Adjust the condenser diaphragm for maximum contrast.*
    a. Switch Optovar ring to PH (phase); this serves as a telescope.
    b. Close condenser diaphragm until the leaves are barely visible.
    c. Focus on the leaves of the diaphragm with the lower ring of the Optovar.
    d. Close the diaphragm 1/3 of the distance of the radius of the visible circle of light.
11. Put the light through the camera by pulling out the beam-splitting slide (to the black ring). The area of the slide to be reproduced is now framed.
12. Center the specimen as desired and adjust the magnification with the Optovar and with the camera magnification projective.
13. REFOCUS THE SPECIMEN.
14. Set the light intensity switch on the automatic exposure device to not more than IX.
15. Set the selector switch on the automatic exposure device to the desired setting. (This switch regulates the photosensitivity of the automatic exposure device.) The setting must be calibrated according to the accompanying instructions and varies with the type of film, the filters, and the individual microscope.
16. Push the shutter button (A) on the automatic exposure device which exposes the film automatically.
17. For multiple photographs, shortcuts must be determined by the individual. It is always wise to center the condenser before each photograph.
18. Remove film cassette from the photomicroscope, and rewind according to the accompanying instructions *before* opening.

GENERAL COMMENTS

This photomicroscope assembly produces excellent results for routine work. Large numbers of photographs may be processed. Focusing through the eyepieces, automatic exposure, and ease of developing and printing 35 mm film account for the increased speed of processing.

---

* Many people who use this photomicroscope close the condenser diaphragm part way to obtain contrast, but cannot reproduce this position from one photograph to another. The use of the PH (phase) ring of the Optovar allows the periphery of the illuminated field (where the light is distorted chromatically) to be cut off by a reproducible amount.

RECORDING OF LOCATIONS

It is important to keep a record of the exact location of a cell or metaphase under study.

Graduated Mechanical Stage

The graduated mechanical stage is the easiest method for recording locations. If the slide must be moved from one microscope to another, both equipped with graduated mechanical stages, a correction factor for the horizontal and another for the vertical position must be calculated.

Field Finder

A ruled and numbered microscope slide may be purchased (Lovins Micro-slide Field Finder, Cat. No. 7100, W. & L. E. Gurley, Troy, N. Y.). This field finder allows one to return to the same position on a microscope with a nongraduated mechanical stage.

1. Locate desired metaphase under the oil immersion objective.
2. Remove slide without changing position of the mechanical stage.
3. Place field finder carefully in position on the mechanical stage. Care must be taken not to chip the edges of the field finder.
4. Record the position marks seen on the field finder—for example, A-5. (40X high dry objective is sufficient magnification.) Also record the slide identification.
5. When the metaphase is to be relocated, the above process is reversed. Place the field finder on the mechanical stage first, locate the position marks (A-5), remove the field finder, and put on the slide with metaphase to be relocated. This metaphase should be in the center of the high dry objective (40X). Switch to oil immersion as needed.

THE CAMERA LUCIDA

The camera lucida attachment projects an image of the metaphase to the side of the microscope where the image may be superimposed on drawing paper. With the proper adjustment of light through the microscope and outside it, a drawing can be made along the projected image on the paper. A Zeiss camera lucida attachment can be added to a standard Zeiss microscope in such a way that it remains

107

in place when not in use. The camera lucida unquestionably facilitates detailed scale drawings. It does not facilitate the quick line drawings used for eye karyotyping.

PHOTOGRAPHY RECORDS

A permanent photography notebook should be kept near the photomicroscope. In this book are noted:

1. The roll and frame number or sheet film code.
2. Slide identification.
3. Location of metaphase or cell.
4. Light intensity setting and selector switch setting (if automatic exposure device is used); seconds of exposure and light meter reading if not automatic.
5. Magnification, including bellows height if indicated.
6. Type of filter, if any.
7. Date of photography.
8. Description (brief) of subject matter of the photograph.
9. Type of film.

For routine photography the entire description need not be listed for each photograph.

MICROSCOPE FILTERS

Filters on the microscope light source are used to add contrast to the photographic image. They also facilitate examination of chromosomes by eye.

Wratten Light Filters (Eastman Kodak Co., Rochester, N. Y.)

Stain	Filter Color
Violet	Yellow
Purple	Green
Blue	Red
Green	Red
Yellow	Blue
Red	Green
Brownish	Blue

FILM

For a more detailed discussion of different types of film, please refer to "Applied Photography in Chromosome Studies" by L. P. Christenson (References, page 94).

35 mm film

Plus-X Panchromatic (Eastman Kodak), which has medium speed, fine grain, and medium-to-good contrast, is a good film for chromosome photography. It is much cheaper to buy 100 ft rolls than individual rolls holding a certain number of frames. Snap-cap magazines and a daylight film loader should also be purchased. Metal film cans for storing the loaded magazines can usually be obtained free of cost at a photography store. A good daylight loader is Watson film loader, Model 66B4 (Burke & James, Inc., Chicago, Ill.). The 100 ft roll is placed in this loader in total darkness, according to the accompanying instructions. In the light, a magazine is inserted, the top is closed, the desired number of frames is rolled on the spool, and the film is cut off. If 20 pictures are desired per roll, about 26 to 27 frames should be rolled. Store extra 100 ft rolls in the refrigerator.

Sheet film (4″ x 5″)

Contrast Process Panchromatic (Eastman Kodak) is good for chromosome photography. It is supplied in boxes of 25 sheets which should be stored in the refrigerator. This sheet film has medium speed, very high contrast, and very high resolving power.

DEVELOPING OF FILM

35 mm film (Plus-X, Panchromatic Eastman Kodak)

A small daylight developing tank for 35 mm film is recommended (Nikor Products Co., Springfield, Mass.). This model consists of a round stainless steel can, removable lid, smaller removable lid through which fluids, but not light, may be introduced, and spools (two fit in one can) to hold the film.

1. In the darkroom in total darkness, remove the exposed film from the snap-cap magazine and roll on the film holder of the daylight developing tank. If two rolls of film are loaded, keep in order

because metaphases are hard to tell apart, short of returning to the slide locations.

2. Place the spool(s) in the developing can; put the lid on tightly. The light may now be turned on.

3. Through the lightproof small opening on the top, pour 1:1 D-11 developer (Eastman Kodak, diluted freshly with an equal volume of tap H_2O), at 70°F. Diluted D-11 is not reused. Stock D-11 is prepared according to the accompanying instructions and stored on a shelf in the dark room, in a dark bottle.

4. Develop for 5 min; agitate for 5 sec, every 30 sec.

5. Pour out the developer and fill with tap H_2O at 70°F.

6. Pour out the tap H_2O and fill with fixer (hypo) at 70°F (Eastman Rapid Fixer with hardener or Edwal Quick Fix). Fixer should be diluted as instructed for negative film. When diluted, it may be reused until "fixer check" fluid (Edwal Hypo-Check) forms a white precipitate. Fix for 5 min.

7. Remove the large lid and run tap H_2O gently into the can. Rinse 30 min.

8. Pour out the tap H_2O and pour in Photo-Flo (Eastman Kodak, diluted as instructed). Pour off. It can be reused. This procedure prevents spotting of the film as it dries.

9. Unwind the film and hang up to dry. Hang by pinch-type clothes pins or individual dental film clips.

10. When dry, write the film number at the beginning of the film.

4" x 5"

A small developing tank for 4" × 5" film is recommended (Yankee, adjustable, film-developing Agitank, Windman Brothers, Los Angeles). This model consists of a rectangular plastic box, removable lid, and adjustable plastic film holder for up to 12 negatives. Fluid capacity for 4" × 5" film is 55 oz.

1. In the darkroom in total darkness, remove the film from the film holders. Place in order in the daylight developing tank film holder.

2. Place the cover on tightly. The light may now be turned on.

3. Through the lightproof opening on the top add D-11 developer (Eastman Kodak) at 70°F. Stock is prepared as instructed. It may be reused several times.

4. Proceed as for 35 mm negatives.

5. Remove the negatives in order, and hang from numbered individual clips to dry.

PRINTING

Polycontrast Rapid paper (Eastman Kodak) is a good all-purpose paper for printing chromosomes. It may be kept at refrigerator temperature for prolonged storage.

35 mm negatives

1. Put 35 mm negative strip in a 35 mm or multipurpose enlarger, emulsion side down.

2. Place a sheet of white 5″ × 7″ paper in the same size easel and turn the enlarger light on it. Adjust the height so that picture to be printed fits the paper.

3. Turn out the overhead lights to focus the image on the paper.

4. With overhead lights on and enlarger light out, set f stop and exposure time. Initially, do not use a polycontrast filter.

5. Turn out overhead lights and leave on yellow safelight (Kodak safelight filter, Wratten Series OC), which may be plugged into the automatic timer on the enlarger. (This safelight then goes out automatically while the enlarger is exposing photographic paper.)

6. Put the photographic paper (5″ × 7″) in the easel, emulsion (slick) side up.

7. Release automatic timer on the enlarger light.

8. Place exposed paper in 1:2 Dektol developer (Eastman Kodak, 1 part Dektol to 2 parts tap H_2O) at 70°F. Diluted Dektol is not reused. Stock Dektol is prepared according to the accompanying instructions and stored on a shelf in the darkroom, in a dark bottle. The proper contrast on the print should appear in about 1 to 1½ min.

9. Move to stop bath for a few seconds (7.5 ml glacial acetic acid in 1 liter of tap H_2O).

10. Fix for 5 min (Eastman Rapid Fixer with hardener or Edwal Quick Fix). Fixer should be diluted as instructed for paper. When diluted, it may be reused until "fixer check" fluid (Edwal

111

Hypo-Check) forms a white precipitate. Lights may be turned on as soon as the paper is in the fixer.

11. Evaluate the print for contrast and intensity. The f opening and exposure time must be adjusted accordingly. To regulate the contrast, polycontrast filters are used. No. 4 filter is highest contrast, and No. 1 is lowest contrast.

12. Wash in tap H_2O at $70°F$ for 1 hr.

13. Soak in Pakosol (Pako Corp., Minneapolis, Minn.) for 5 to 10 min. This process conditions the prints and makes them easier to flatten.

14. Place on print dryer (Omega color print dryer is good; two sizes are available).

4″ x 5″ negatives

1. Follow instructions for the enlarger. Lenses, condensers, and auto-focus tract will need adjustment.

2. Be sure to insert the film with emulsion side down.

3. Determine f opening, exposure time, and use of polycontrast filters as for 35 mm negatives.

4. Process prints as already described for 35 mm.

OTHER TYPES OF LIGHT MICROSCOPY

(Fluorescence, near-ultraviolet, ultraviolet, dark-field, phase contrast, interference, polarizing, oblique incident lighting)

References: Leitz and Zeiss instruction manuals for phase and fluorescent microscopy.

Needham, G. H.: *The Practical Use of the Microscope.* Springfield, Ill., Charles C Thomas, 1958, Ch. VIII, The Fluorescence Microscope; Ch. IX, The Near Ultraviolet Microscope; the Ultraviolet Microscope; the Reflecting Microscope; Ch. XI, The Phase Contrast Microscope; the Interference Microscope; Ch. XII, The Polarizing Microscope.

Razavi, L.: Single Stranded DNA in Lymphocyte Chromosomes (Immunofluorescence Technique). Nature, *215:*928, 1967.

Runge, W. J.: Bright Field, Phase Contrast, and Fluorescent Microscopy. In: *Human Chromosome Methodology,* J. J. Yunis, ed. New York, Academic Press, 1965.

Schneider, L. E.: The Microscopic, Three-Dimensional Viewing of Mammalian Chromosomes. Exp. Cell Res., *47:*658, 1967.

Sharma, A. K., and Sharma, A.: *Chromosome Techniques. Theory and Practice.* Washington, Butterworths, 1965, Ch. 9.

ELECTRON MICROSCOPY

References: Brinkley, B. R., Murphy, P., and Richardson, L. C.: Procedure for Embedding *in Situ* Selected Cells Cultured *in Vitro*. J. Cell Biol., *35:*279, 1967.

Caro, L. G.: High-Resolution Autoradiography. In: *Methods in Cell Physiology,* Vol. I, D. M. Prescott, ed. New York, Academic Press, 1964, Ch. 16.

Christenhuss, R., Buchner, T., and Pfeiffer, R. A.: Visualization of Human Somatic Chromosomes by Scanning Electron Microscopy. Nature, *216:*379, 1967.

DuPraw, E. J.: Evidence for a "Folded-Fibre" Organization in Human Chromosomes. Nature, *209:*577, 1966.

Hoskins, G. C.: Sensitivity of Microsurgically Removed Chromosomal Spindle Fibers to Enzyme Disruption. Nature, *217:*748, 1968.

Needham, G. H.: *The Practical Use of the Microscope.* Springfield, Ill., Charles C Thomas, 1958, Ch. X. The Electron Microscope.

Osgood, E. E., Jenkins, D. P., Brooks, R., and Lawson, R. K.: Electron Micrographic Studies of the Expanded and Uncoiled Chromosomes from Human Lymphocytes. Ann. N. Y. Acad. Sci., *113:*717, 1964.

Wolfe, S. L.: The Fine Structure of Isolated Metaphase Chromosomes. Exp. Cell Res., *37:*45, 1965.

113

9

Light Microscope Tritium Autoradiography

The *tritiated thymidine* labeling patterns at the end of the *synthesis period (S)* are important for the identification of normal and abnormal human chromosomes (see Chapter 3, Tables 3-7a and 3-7b). *Replication* patterns during other portions of the S period are not as well standardized, except for X chromosomes in addition to one, which are clearly late to start and late to finish replication (Figs. 9-1 and 9-2).

This chapter covers methods to prepare cells for *autoradiography,* and the application and processing of photographic films. The technique is important, not only to study chromosome labeling patterns but also to study the labeling of *sex chromatin,* events of the *cell cycle* (see Chapter 15 covering special procedures on cells in culture), and mechanisms of *DNA* replication.

The physical properties of tritium are described in Table 9-1. Because the emissions are weak β-particles (average path length 1 μ, maximum travel distance 6 μ) and because the half-life is 12.4 years, tritium is well suited for the procedures described in this chapter. Disposal is not difficult, and the short emission path can be used to expose a thin photographic film applied very close to the specimen.

114

Table 9-1. Physical Properties of Tritium

Atomic weight	3
Half-life	12.4 years
Mode of radioactive decay	β, no γ
Product of decay	He^3
Maximum energy, β-ray	18 kev
Average energy, β-ray	5.7 kev

References: Bender, M. A., and Prescott, D. M.: DNA Synthesis and Mitosis in Cultures of Human Peripheral Leukocytes. Exp. Cell Res., *27:*221, 1962.

Caro, L. G., and van Tubergen, R. P.: High-Resolution Autoradiography. I. Methods. J. Cell Biol., *15:*173, 1962.

Kopriwa, B. M., and Leblond, L. P.: Improvements in the Coating Technique of Radioautography. J. Histochem. Cytochem., *10:*269, 1962.

Perry, R. P.: Quantitative Autoradiography. In: *Methods in Cell Physiology*, Vol. I. D. M. Prescott, ed. New York, Academic Press, 1964, Ch. 15.

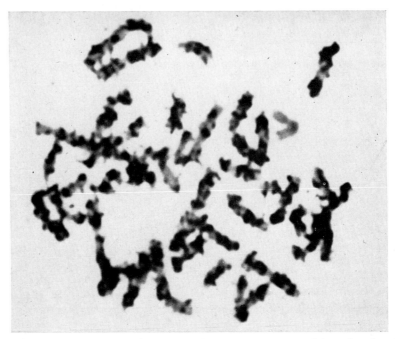

Figure 9-1. *Autoradiograph of metaphase from a normal female, showing an unlabeled X chromosome at the beginning of S. The other chromosomes are heavily labeled. This cell in long term culture was treated with FUdR and labeled with H_3TdR for the first 2 hr of S. Ilford L-4 emulsion. Giemsa stain.*

Rogers, A. W.: *Techniques of Autoradiography*. New York, Elsevier Publishing Co., 1967.

Schmid, W.: Autoradiography of Human Chromosomes. In: *Human Chromosome Methodology*, J. J. Yunis, ed. New York, Academic Press, 1965.

Stubblefield, E.: Quantitative Tritium Autoradiography of Mammalian Chromosomes. I. The Basic Method. J. Cell Biol., *25:*137, 1965.

Wimber, D. E.: Effects of Intracellular Irradiation with Tritium. Advances Rad. Biol., *1:*85, 1964.

PREPARING THE SPECIMEN FOR AUTORADIOGRAPHY

Reference: Prescott, D. M., and Bender, M. A.: Preparation of Mammalian Metaphase Chromosomes for Autoradiography. In: *Methods in Cell Physiology*, Vol. I. New York, Academic Press, 1964, Ch. 19.

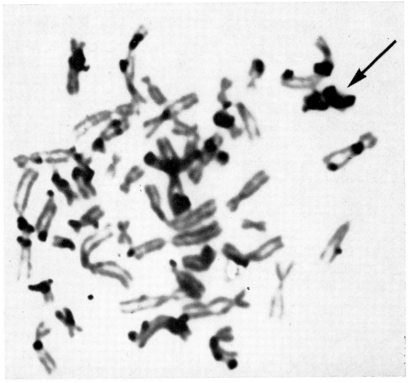

Figure 9-2. *Autoradiograph of metaphase from a normal female, showing a heavily labeled X chromosome at the end of S (arrow). This cell in long term culture was labeled for 6 hr with H₃TdR. During the last 3 hr of this time colchicine was applied. NTB-3 emulsion. Giemsa stain.*

Peripheral Blood

Chromosome Labeling Patterns toward End of S

LABELING PROCEDURE

Prepare a short-term culture of human peripheral blood as described in Chapter 4. Be familiar with handling and disposal procedures for tritium described later in this section. Use medium without TdR.

a. For demonstration of X chromosomes which are late to finish replication (Fig. 9-2), incubate the cultures for 72 to 85 hr after the addition of phytohemagglutinin. Add H_3TdR (specific activity, 0.36 curies/mM, Schwartz BioResearch, Inc., Orangeburg, N. Y.) to each culture in a final concentration of 1.0

μc/ml, at 37°C. Continue incubation 1 hr. Add colchicine (final concentration 5×10^{-6} M). Continue incubation 8 to 9 hr after the addition of H_3TdR.

Method of: Mukherjee, B. B., Miller, O. J., Breg, W. R., and Bader, S.: Chromosome Duplication in Cultured Leukocytes from Presumptive XXX and XXXXY Human Subjects. Exp. Cell Res., *34:*333, 1964.

b. For demonstration of terminal chromosome labeling patterns, a shorter H_3TdR labeling time is indicated. Incubate the cultures for approximately 72 hr. Add H_3TdR (specific activity 1.9 c/mM) to each culture in a final concentration of 1.0 μc/ml 6 hr prior to termination of the cultures. Two hr before termination of the cultures, add Colcemid (Ciba) to each culture in a final concentration of 0.03 μg/ml. Since the G_2 period (gap period between the S period and mitosis) in the cell cycle of human cells in leukocyte culture is about 4 hr, the actual time span during which the chromosomes incorporate radioactive thymidine in this system ranges from a few minutes up to about 2 hr. It is important to maintain temperature at 37°C during the entire labeling time.

Method of: Schmid, W.: DNA Replication Patterns of Human Chromosomes. Cytogenetics, *2:*175, 1963.

or

Modifications of b. above

1. H_3TdR is added for the last 6 hr and Colcemid for the last 3 hr.

Method of: German, J.: The Pattern of DNA Synthesis in the Chromosomes of Human Blood Cells. J. Cell Biol., *20:*37, 1964.

2. Label the cells for 3 hr prior to processing and apply Colcemid for 1.5 hr prior to processing.

Method of: Gilbert, C. W., Muldal, S., and Lajtha, L. G.: Rate of Chromosome Duplication at the End of the Deoxyribonucleic Acid Synthetic Period in Human Blood Cells. Nature, *208:*159, 1965.

PREPARATION OF CHROMOSOMES

Because radioactivity is present in the medium, certain modifications are introduced into the routine processing of peripheral blood cells.

1. Pour medium and cells into 12-ml heavy duty graduated centrifuge tube. Centrifuge at 600 to 800 rpm for about 8 min.

2. Pour supernatant carefully down sink with running tap H_2O. Aspirate last drops with a Pasteur pipette and wash these drops down the sink with running H_2O. (Note: the handling of tritium depends on the amounts used. If a large amount is used, disposal procedures other than the one described may be prescribed by the Radiation Safety Committee or other responsible committee of the institution.)

3. Add several ml of balanced salt solution (PBS, see Chapter 12), suspend cells gently, and centrifuge as before.

4. Remove all of the supernatant as described under 2.

5. Add 0.5 ml PBS. Add to 2.0 mark with distilled H_2O. Suspend cells gently. Incubate at 37°C for 10 min. (Some investigators leave at room temperature.)

6. Shake gently and centrifuge as before. Aspirate all supernatant as close to the top of the button as possible. Discard with running H_2O as described under 2.

7. Add 1 to 2 ml of 1:3 fixative (1 part glacial acetic acid to 3 parts absolute methanol) carefully down side of tube, without disturbing the button of cells.

8. Allow to stand at room temperature for 30 min.

9. Resuspend cells by pushing out with Pasteur pipette. Centrifuge as before.

10. Add about 0.2 ml 45% acetic acid to make a dense cell suspension and prepare air-dried slides as described in Chapter 7. Slides should be precleaned by soaking in 95% ethanol and drying well with lint-free cloth.

or

Add 1:3 fixative, suspend, centrifuge, decant, repeat fixative change, and add about 0.2 ml fresh 1:3 fixative. Prepare flame-dried slides as described in Chapter 7.

or

Make squash preparations (Chapter 7).

11. For chromosome autoradiography well spread metaphases of excellent quality are *essential*. It is wise to prepare as many slides as possible, since some may be lost through technical problems with the autoradiography (accidental exposure of the film, etc.).

119

PREPARATION OF SLIDES FOR AUTORADIOGRAPHY

Rinse the air- and flame-dried slides in gently running tap H_2O for at least 4 hr prior to the application of emulsion. Air dry. Never apply emulsion to all the slides of one experiment at one time unless the experiment can be easily repeated.

Squash slides require a different preparation for the application of photographic film. (See Chapter 7 for further details of squash preparations.)

1. Place temporary squash preparation slides, right side up, on the smooth surface of a block of dry ice.
2. Leave for about 5 min, during which time the wax edges will crack.
3. With a scalpel, flip the coverglass off with a single, quick movement.
4. Place the slide in a staining jar containing absolute alcohol.
5. After 4 min remove the remnants of cement (used for temporary mounting) with a razor blade and place the slides for 2 min in another jar containing absolute alcohol. Careless handling during this procedure will cause the loss of many cells.
6. Place the slides in a slide box and dry. They may be stored for a long period without deteriorating.

Method of: Schmid, W.: DNA Replication Patterns of Human Chromosomes. Cytogenetics, *2:*175, 1963.

EXPOSURE OF THE PHOTOGRAPHIC FILM

Expose for the times described below (a general guide):

AR-10 stripping film	4 to 7 days
NTB-2	2 to 7 days
NTB-3	2 to 7 days
Ilford L-4	2 to 3 weeks (or longer)

See Table 9-2 (page 129) and other sections later in this chapter for further consideration of the different types of photographic emulsion.

Chromosome Labeling Patterns toward Beginning of S

1. Prepare a short-term culture of human peripheral blood as described in Chapter 4. Be familiar with handling and disposal procedures for tritium. Use medium without TdR.

120

2. At the 20th hr of incubation, add H_3TdR (1 μc/ml, specific activity 6.7 c/mM, New England Nuclear Corp.).

3. After 10 hr of incubation with H_3TdR, add TdR to 100$\times$ that of the H_3TdR.

4. Transfer the culture to a centrifuge tube and spin gently. Discard the radioactive supernatant with running tap H_2O.

5. Resuspend in fresh medium (37°C) containing TdR to 100 $\times$, as above. (A second rinse in 100$\times$ TdR medium may be indicated.)

6. Continue incubation and add colchicine to the culture from the 44th to the 47th hr (or 43rd to 53rd hr).

7. Process as described in this chapter for studying the chromosome labeling patterns in peripheral blood toward the end of S (page 118).

Method of: Takagi, N., and Sandberg, A. A.: Chronology and Pattern of Human Chromosome Replication. VIII. Behavior of the X and Y in Early S-Phase. Cytogenetics, 7:135, 1968.

Chromosome Labeling Patterns at Intervals during S

1. Prepare a short-term culture of human peripheral blood as described in Chapter 4. Be familiar with handling and disposal procedures for tritium. Use medium without TdR.

2. At the 70th hr of incubation, add H_3TdR (1 μc/ml, specific activity 6.7 c/mM, New England Nuclear Corp.) for 10 min. Before labeling, save half the culture medium for use in a "chase" incubation with fresh medium (semi-conditioned medium, step 6). Exert every effort to avoid temperature change during manipulation of the culture.

3. After 10 min of incubation with H_3TdR, add TdR to 100$\times$ that of the H_3TdR.

4. Transfer the culture to a centrifuge tube and spin gently.

5. Discard the radioactive supernatant with running tap H_2O and substitute with fresh medium. Resuspend.

6. Centrifuge as before and reincubate the cells in semi-conditioned culture medium containing TdR to 100$\times$, as above.

7. Remove 8 ml aliquots of cell suspension from the culture at hourly intervals (over 13 hr) after labeling. To each aliquot add 0.2 ml of 0.05% colchicine.

121

8. Incubate each aliquot for 1 hr and process immediately as described in this chapter for studying the chromosome labeling patterns in peripheral blood toward the end of S (page 118).

Method of: Takagi, N., and Sandberg, A. A.: Chronology and Pattern of Human Chromosome Replication. VII. Cellular and Chromosomal DNA Behavior. Cytogenetics, 7:118, 1968.

Monolayer Cells in Long-term Culture

Chromosome Labeling Patterns toward End of S

1. Prepare replicate monolayer cultures as described in Chapter 15, "Replicate Plating Protocol." Be familiar with handling and disposal procedures for tritium. Use medium without TdR.

2. To cultures in logarithmic growth (not confluent), add H_3TdR (0.36 c/mM) to a final concentration of 0.1 μc/ml 4 to 6 hr prior to termination of the cultures.

3. Apply colchicine to a final concentration of 0.5 μgm/ml (or 0.1 μgm/ml) for the last 3 hr of label.

4. Process for chromosomes as described in Chapter 5, "Preparation of Chromosomes from Human Cells in Long-term Culture." Be sure to discard radioactive solutions down the sink with running H_2O.

Method of: Priest, J. H.: Unpublished.

or

Modification of 2. and 3. above.

Add H_3TdR (New England Nuclear, 2 c/mM), 1 μc/ml final concentration in the culture medium for 15 min at 37°C. Remove medium containing isotope and replace by fresh medium containing 0.01 mg/ml of TdR. Repeat this medium change and reincubate 4 hr prior to processing. Add colchicine for the final 1½ to 2 hr of incubation.

Method of: Atkins, L., Book, J. A., Gustavson, K. H., Hansson, O., and Hjelm, M.: A Case of XXXXY Sex Chromosome Anomaly with Autoradiographic Studies. Cytogenetics, 2:208, 1963.

Chromosome Labeling Patterns at Beginning of S
(FUdR Synchronized Cells)

1. Prepare replicate monolayer cultures as described in Chapter 15, "Replicate Plating Protocol." Be familiar with handling and disposal procedures for tritium. Use medium without TdR.

2. Add FUdR (5-fluoro-2'-deoxyuridine, Hoffman-La Roche, Inc., Nutley, N. J., Research Division) to a final concentration of 0.1 μgm/ml (4×10^{-7} M). Also add uridine 6×10^{-6} M. Continue incubation for 16 to 20 hr. (This synchronization procedure is also described in Chapter 15.)

3. Pour off medium containing FUdR and add culture medium containing H_3TdR (0.36 c/mM, Schwartz BioResearch, Inc.). The length of the labeling time depends on the design of the experiment. To study the 1st 10 min of S, label for 10 min; to study the 1st 20 min, label for 20 min; and so forth up to 4 to 6 hr labeling periods. Use the following amounts of H_3TdR:

Length of Label	μc of H_3TdR (final concentration/ml)
10 to 30 min	2 to 4
1 to 2 hr	0.5
3 hr	0.2
4 to 5 hr	0.1

4. To terminate pulse, pour off radioactive medium, rinse twice with balanced salt solution (PBS, see Chapter 12), and add medium containing 6×10^{-6}M TdR.

5. Five hr after onset of label, add colchicine to a final concentration of 0.5 μgm/ml (or 0.1 μgm/ml) or vinblastine sulfate (Velban) to a final concentration of 0.1 μgm/ml.

6. Five hr later (10 hr after onset of label), process the slides for chromosomes as described in Chapter 5, "Preparation of Chromosomes from Cells in Long-term Culture." Be sure to discard radioactive solutions down the sink with running H_2O, taking the precautions already described in this chapter for studying chromosome labeling patterns in peripheral blood (page 118).

Method of: Priest, J. H., Heady, J. E., and Priest, R. E.: Delayed Onset of Replication of Human X Chromosomes. J. Cell Biol., *35:*483, 1967.

Chromosome Labeling Patterns at Intervals during S

1. Prepare replicate monolayer cultures as described in Chapter 15, "Replicate Plating Protocol." Be familiar with handling and disposal procedures for tritium. Use medium without TdR.

2. To cultures in logarithmic growth (not confluent), add H_3TdR (0.36 c/mM) 4 μc/ml final concentration. Incubate 30 min.

3. Two hr after onset of label, add vinblastine sulfate (Velban) 0.1 μgm/ml final concentration for 1 hr to the first set of cells (specimen 1) and process immediately for chromosomes as described in Chapter 5, "Preparation of Chromosomes from Cells in Long-term Culture." Be sure to discard radioactive solutions down the sink with running H_2O, taking the precautions already described in this chapter for studying chromosome labeling patterns in peripheral blood (page 118).

4. Subsequently process sets of cells (specimens 2 to 11) each hr, always after 1 hr of Velban according to the following plan:

Sampling hr	0	1	2	3	4	5	6	7	8	9	10	11	12	13
		G_2				S							G_1	
Cell cycle hr		4	3	2	1	8	7	6	5	4	3	2	1	
Specimen No.				1	2	3	4	5	6	7	8	9	10	11

5. Since this is an extremely long experiment, it may be modified by selecting the parts of S that are of interest, or (less ideally) if all of S is needed, the experiment may be divided in half. An overlapping interval in the middle should be run at both times as a control.

6. Interpretation: Specimen No. 1 should have very few labeled metaphases since the mitotic cells on the slides were in G_2 at the time of application of H_3TdR. A few more labeled metaphases will appear in specimen No. 2. Replication patterns in the last 2 hr of S should be present in labeled metaphases in specimen Nos. 3 and 4. This is an excellent time to demonstrate late replicating X chromosomes (Fig. 9-2). Specimen Nos. 5 through 7 show labeled metaphases with chromosome replication patterns characteristic of the middle S. There is reasonably uniform labeling over all the chromosomes. Specimen Nos. 8 through 10 show unlabeled X chromosomes late to start replication (Fig. 9-1); labeled metaphases are characteristic of the first

part of S. Specimens after No. 10 should show unlabeled metaphases, since the mitotic cells on the slides were in G_1 at the time of application of H_3TdR.

Method of: Priest, J. H.: The Replication of Human Heterochromatin in Serial Culture. Chromosoma, *24:*438, 1968.

Sex Chromatin Labeling Patterns during S (FUdR Synchronized Cells) (Fig. 9-3)

LABELING PROCEDURE AND PREPARATION OF SLIDES

1. Prepare replicate monolayer cultures as described in Chapter 15. Be familiar with handling and disposal procedures for tritium. Use medium without TdR.

2. Apply the FUdR synchronization technique as described in this chapter under chromosome labeling patterns at the begin ning of S in long-term culture (page 123).

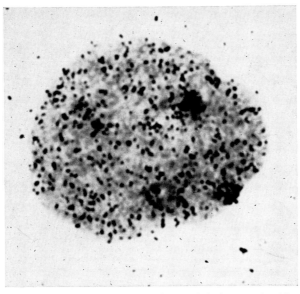

Figure 9-3. *Autoradiograph showing heavy label over double sex chromatin bodies in an interphase from a triple X female. This cell in long-term culture was treated with FUdR and labeled towards the end of S with H_3TdR. Heavy label over sex chromatin is a pattern characteristic of the last several hr of S. Ilford L-4 emulsion. Giemsa stain.*

3. To terminate the FUdR block and label the cells, pour off medium
    containing FUdR and add culture medium containing H₃TdR
    (0.36 c/mM, Schwartz BioResearch, Inc.). The length of the
    labeling time depends on the interval of S to be studied. To
    study the 1st 10 min of S, label for 10 min; to study the 1st 20
    min, label for 20 min; and so forth. By this technique a con-
    tinuous label accumulates during the interval to be studied. Use
    the following amounts of H₃TdR:

Length of Label	*μc of H_3TdR (final concentration/ml)*
10 to 30 min	2 to 4
1 to 2 hr	0.5
3 hr	0.2

    An alternative labeling technique is to apply 10-min pulses of
    H₃TdR (4 μc/ml final concentration) at intervals during S.
    The first 10-min pulse is applied as above. For all the other
    specimens, pour off medium containing FUdR and add medium
    containing 6×10^{-6} M TdR to reverse the block without labeling
    the cells. At 30-min intervals thereafter, depending on the inter-
    vals to be studied, pour off the TdR medium from the appropri-
    ate specimens, rinse 2 times with balanced salt solution (or
    medium without TdR), after each rinse aspirate the last drops
    of fluid and discard, and add H₃TdR medium. Temperature
    must be maintained at 37°C. Incubate for 10 min.

4. Termination of the label and onset of processing of the cells
    for sex chromatin are both accomplished at the same time. Pour
    off medium and discard down the sink with running tap H_2O.
    Aspirate last drops and discard with running H_2O.

5. Add balanced salt solution (PBS, Chapter 12), rinse, pour off
    with running H_2O, aspirate last drops as above. Repeat this
    rinse.

6. Add 0.5 to 1 ml 0.25% trypsin-0.25% Versene in PBS (Chapter
    12) and incubate 1-plus min until cells come off.

7. Add PBS and centrifuge cell suspension at 600 to 800 rpm for
    about 8 min. Discard supernatant with running H_2O and aspirate
    last drops.

8. Add about 2 ml of fixative (1 part glacial acetic acid to 3 parts
    absolute methanol, freshly made) down side of tube without
    disturbing the button of cells. Fix 30 min.

9. Resuspend the cells, centrifuge, discard the supernatant, and
    aspirate last drops. Repeat.

126

10. Add about 0.2 ml 45% acetic acid (depending on the size of the button).

11. Prepare air-dried slides* by allowing a drop of cell suspension to run down a tilted slide, precleaned in 95% ethanol. Air dry quickly. More than one drop of cell suspension may be placed side by side on the slide.

12. The thionin sex chromatin stain is described here. (Also see Chapter 10, pages 145 and 146.) Include a known sex chromatin positive control slide with each staining.

13. Hydrolyze in 5 N HCl for 20 min at room temperature.

14. Rinse 4 times in distilled H_2O to be sure no acid is carried into thionin stain.

15. Stain in freshly prepared thionin working solution at pH 5.7 (see Chapter 10, page 145) for 10 min (or more).

16. Rinse quickly in distilled H_2O. Air dry.

AUTORADIOGRAPHY

1. If the technique of double photography is used (or single photography, plus a second analysis by eye), appropriate cells should be located, the positions recorded, and photographs made (see Chapter 8).

2. Clean the slides as described later in this chapter. (Procedure for Putting Emulsion on Slides Covered with Immersion Oil, (page 136).

3. Apply NTB-2 or Ilford L-4 emulsion, expose, and develop as described in this chapter.

4. Stain for about 5 min in Giemsa (Chapter 7, page 89).

5. Locate the same cells and analyze the relation of the label to the sex chromatin (or other heterochromatin) in the initial photograph.

6. Take a second photograph as needed (Fig. 9-3 and Chapter 8).

Method of: Priest, J. H.: The Replication of Human Heterochromatin in Serial Culture. Chromosoma, *24:*438, 1968.

---

* Cells air dried from suspension are not as flat as cells grown on coverslips; for sex chromatin studies, flat cells are desirable. However, the coverslip method, which produces flatter cells, has not been successful in this laboratory when cell synchronization (FUdR) is employed. Cells on coverslips are not uniformly in logarithmic growth for application of the block to DNA synthesis.

Labeling of Interphase Nuclei during S

This procedure for labeling interphase nuclei during S is similar to that just described for sex chromatin labeling patterns during S. Synchronization of the cells is not necessary. Labeling of interphase nuclei with H_3TdR is also described in Chapter 15 in the section on the length of the S period (page 208).

Preparation of Autoradiographs from Coverglass Cultures

1. Grow monolayer cultures on Leighton tube coverslips. (See Chapter 6, "Short-term Cultures from Small Tissue Biopsies, Coverglass Method.")
2. Expose cultures to H_3TdR (or H_3uridine, or H_3arginine), according to the rules already described in this chapter.
3. Fix in formalin or acetic alcohol, or follow chromosome procedure described in Chapter 6, page 78, for short-term cultures from small tissue biopsies, coverglass method.
4. Warm paraffin (Fisher Scientific Co.) in a beaker at 52°C.
5. Dip a preflamed slide in the melted paraffin and place it flat on a filter paper.
6. Place the coverglass, cell side up, on the slide and let it cool for 30 sec before immersing in a Coplin jar containing distilled H_2O.
7. Follow routine autoradiographic procedure using either liquid emulsion or stripping film (pages 129-134).
8. Stain the autoradiographs and dehydrate in graded series of alcohol.
9. After a rinse in absolute alcohol, make a clean tear with a sharp blade along the edges of the coverglass and apply a mild flame to the back of the slide.
10. Place the slide quickly in a Petri dish containing xylol to detach the coverglass.
11. Leave the coverglass in xylol until the paraffin dissolves.
12. Rinse the coverglass in fresh xylol and mount on a clean slide in a drop of Permount.

Method of: Basrur, P. K., Basrur, V. R., and Gilman, J. P. W.: Method for the Preparation of Autoradiographs with Coverslip Cultures. Nature, *212:*424, 1966.

LIQUID EMULSION

The grain diameter and relative sensitivity of different emulsions are described in Table 9-2. We have compared NTB-2, NTB-3, and Ilford L-4, and these liquid emulsions are considered in this section. (AR-10 stripping film is considered in the next section, "Stripping Film.")

Ilford L-4 has a fine grain and is useful for studying chromosome labeling patterns when grain counts are not necessary and when double photography is to be avoided, say if a very large number of metaphases are to be analyzed. The grains are irregular in size and therefore difficult to count after a 3-week exposure. Longer exposure times or modification of the developing procedure may make the grains possible to count. Dilution 1 to 1 with H_2O is recommended to facilitate applica-

Table 9-2. Grain Diameter and Relative Sensitivity of Different Emulsions

Emulsion	Grain Diameter	Relative Sensitivity
	μ	
Ilford L-4	0.12	132
Kodak AR-10	—	57
Kodak NTB-3	0.23	48

tion. Ilford L-4 tends to have low background if the emulsion is applied with the usual precautions described in this chapter.

NTB-2 and NTB-3 have the same grain size grossly, which is larger and more regular than Ilford L-4. Less exposure time is required than for Ilford L-4. Dilution (2 parts of emulsion to 3 parts of H_2O) is recommended to facilitate application. NTB-3 tends to have high background in spite of every possible precaution. Some lots of NTB-3 may have an unacceptable amount of background when just purchased. However, NTB-2 tends to have low background if it is applied with the usual precautions.

Reference: Prescott, D. M.: Autoradiography with Liquid Emulsion. In: *Methods in Cell Physiology,* Vol. I. New York, Academic Press, 1964, Ch. 17.

129

Test Slides and Background

Every time emulsion is applied to a specimen, it should also be applied in the same way to a test slide, which is (1) a plain slide precleaned by the same routine as the specimen slide, or (2) an extra slide with labeled cells on it, preferably similar to the cells to be studied. Every time a new lot of emulsion is purchased, an undiluted test slide should be made. Test slides are especially critical near the expiration time of the emulsion or if the undiluted emulsion has been melted, placed in aliquots, and saved. (We do not routinely aliquot undiluted emulsion, especially NTB-3.) It is wise to keep a record book of grain counts on test slides. This book permits each laboratory to set standards for acceptable and unacceptable background. To some extent these standards must be set for each type of experiment. In general, for undiluted NTB-2 and NTB-3 grain counts over 100 grains per oil field are not acceptable. This figure may be higher for Ilford L-4, depending on the experiment. A test slide may be exposed overnight, or longer, depending on how soon the test results are needed.

Application

1. Apply in darkroom, safelight with 10 watt bulb and Kodak Wratten filter No. 2 at a distance of 3 ft.*

2. Remove a small aliquot of emulsion from the main supply and melt in H_2O bath at 43°C (National Appliance, round, 11″ × 5″ OD, disconnect indicator light). Mix 2 parts NTB-2 or NTB-3 with 3 parts distilled H_2O warmed in the same H_2O bath. Mix equal parts of Ilford L-4 and distilled H_2O.

3. Pour diluted emulsion into a plastic 2-slide holder in a beaker or rack in the H_2O bath. This type of holder requires 25 ml of liquid (to the top). A Coplin jar or a larger slide holder may be used, but these require much more liquid to fill them to the proper depth. Since diluted emulsion cannot be reused, it is wise to choose a dipping container that holds an appropriate

---

* Some investigators apply NTB-3 in complete darkness, since this emulsion is extremely light sensitive. We have had good luck with the standard use of a Wratten filter No. 2 for different types of liquid emulsion, including NTB-2, NTB-3, and Ilford L-4. It is extremely important to have no light leaks in the room used for the application of emulsion. Controlled humidity in the room is not important for liquid emulsion.

amount of liquid. These containers must be *clean* before use. They cannot have old emulsion in them.

4. Dip slides in and out, to the level of the label on the slide. This label also serves to mark the side with the cells and emulsion. Wipe off the back and allow to dry in the upright position at room temperature for about 30 min.

5. Pack slides in plastic dark-tight boxes containing dehydrating agent (silica gel) wrapped in gauze or microscope paper. Seal boxes with tape, and store at room temperature (on a shelf in the darkroom). Some investigators recommend storage at refrigerator temperature. It is a good idea to store slides with similar exposure times in the same box. On the outside of the box mark the exact description of the slides inside.

6. Expose the appropriate amount of time:

NTB-2	2 to 7 days
NTB-3	2 to 7 days
Ilford L-4	2 to 3 weeks (or longer)

Developing

1. Place in a slide carrier in D-19 developer for 2 min at 70°F.
2. Place gently in tap H_2O for 30 sec at 70°F.
3. Place in Rapid Fixer (Eastman Kodak) with hardener* for 2 min at 70°F.
4. Place gently in tap H_2O and remove.
5. Place in hypo clearing agent (diluted as instructed) for 1½ min.
6. Place in gently running cold tap H_2O for 2½ min. Cold H_2O keeps the emulsion harder and prevents it from washing off.
7. Air dry or place directly in aqueous stain.

Staining

In our experience the best all-purpose chromosome stain on top of liquid emulsion is Giemsa. It should be buffered with phosphate buffer, pH 6.4 (see Chapter 7, page 89).

1. Stain air-dried slides or slides directly from final H_2O rinse, for 2 to 10 min, depending on the type of specimen.

---

* Some investigators leave out the hardener because they think it tends to crack the emulsion. We leave hardener in because without it the emulsion has a tendency to wash off.

2. Rinse gently in distilled H_2O and dry.

3. Thoroughly dry slides may be mounted from xylene, or they may be left unmounted. We have found unmounted slides easier to work with. They can be destained and restained. Immersion oil may be placed directly on thoroughly dry emulsion. Do not wipe the oil off. For cleaning, place the slide directly in two changes of xylene for about 1 to 2 min each time. Blot dry gently between bibulous paper.

4. If needed, destain by placing in xylene (to remove oil as already described), absolute ethanol, and 95% ethanol and leave in 70% ethanol about 1 to 2 min. Place in 95% ethanol; air dry. If a darker stain is needed, place in xylene (to remove oil as already described). Blot dry gently and then air dry. Place in Giemsa for about 5 more min.

Chromosome stains that involve hydrolysis (Feulgen, thionin) may not be used on top of emulsion because it is partially removed by hydrolysis. Stains in acetic acid (orcein) cannot be used. Prescott (see reference, page 129) recommends aqueous solution of toluidine blue (0.25% w/v) at about pH 6. Excess toluidine blue is washed off with 95% alcohol, and the slides are air dried.

Sex chromatin stains after emulsion have not been satisfactory in this laboratory. Giemsa will stain sex chromatin, but it also stains nucleoli. For sex chromatin autoradiography a sex chromatin stain (thionin, Chapter 10), is performed before the application of emulsion. Cells are photographed, emulsion is applied, film is exposed and developed, and the slides are restained with Giemsa. The same labeled cells are relocated and analyzed or photographed. (Also see section, this chapter, on sex chromatin labeling patterns during S.)

STRIPPING FILM

Some laboratories prefer to use stripping film (Kodak AR-10) instead of liquid emulsion. The choice depends, to some extent, on which technique is worked out best in the individual laboratory. In the hands of an experienced person stripping film is not more difficult to apply (as is frequently said). Emulsion is always uneven and thicker at the end of the slide that is lower during drying. There is never this problem with stripping film.

Application

1. Cut film, mounted on glass plates, into squares of approximately 40 $\times$ 40 mm. In order to keep background grains to a minimum, perform the cutting by using a frame on top of which a ruler can be moved without touching the film. Use a scalpel for cutting, and a new blade for each film plate.

2. Place the plate into a tray containing 75% alcohol. Transfer after 3 min to a 2nd tray containing absolute alcohol.

3. With small forceps, lift a single film square and remove from the plate. Drop, emulsion side down, into a 3rd tray containing distilled H_2O at room temperature. Film should spread out completely on the surface of the H_2O.

4. With one hand dip a slide in the water, and with the other hand maneuver the film square into position with one end of the film touching the slide near its labeled end.

5. Lift the slide out of the H_2O. The film will cover the specimen, and the overlapping ends of the film will fold to the underside of the slide.

6. Place the slide on a glass plate and attach with a clamp.

7. Place plate, with about 6 slides on it, back into the tray with distilled H_2O for about 5 min.

8. Remove plate from H_2O and slides from the plate, one by one. Straighten out the film square with the aid of a camel hair brush and the fingers.

9. Place in drying rack and air dry at room temperature in dark room or by an electric blower in a dark-tight box.

10. When slides are dry, place in a small slide box, with about a teaspoonful of dehydrating agent (silica gel), wrapped in gauze or microscope paper. Tape around the lid of the box to prevent light leakage.

11. Place box in the refrigerator until the slides are ready to be developed. Expose 4 to 7 days.

Developing

1. Paint the reverse sides of the slides with mounting medium (Permount, Euparal) in order to prevent shifting of the film squares.

133

2. Dry as already described under application.
3. Place slides in a 10- or 20-slide rack to facilitate moving them simultaneously from one fluid to another.
4. Place for 2 min in D-19B (Eastman Kodak) at 20°C. (5 min is too long; 2 min gives smaller grains.)
5. Dip the slides twice in water.
6. Fix in acid fixer for twice the time required for the film to clear, or approximately 2 min.
7. Wash in running H_2O bath at 20°C for 5 to 10 min.
8. Air dry or use drying box already described.

Staining

1. Stain in Giemsa for about 7 min (see Chapter 7, page 89).
2. Dip twice in distilled H_2O and air dry.
3. Use a razor blade to remove the overlapping ends of the film on the back of the slides.
4. Mount dry slides, or leave unmounted.

Method of: Schmid, W.: DNA Replication Patterns of Human Chromo-somes. Cytogenetics, *2:*175, 1963.

DOUBLE PHOTOGRAPHY

Double photography permits analysis of chromosome or sex chromatin labeling patterns by a comparison of both labeled and unlabeled photographs.

Method 1

1. Apply emulsion as described in this chapter. Expose, develop, and stain.
2. Locate the desired cells and photograph as described in Chapter 8. Be sure to keep an exact record of the locations.
3. Remove the silver grains as described in this chapter, under special procedures. Restain (with Giemsa, see Chapter 7).
4. Relocate the same cells and photograph.

Method 2

1. Locate the desired cells and photograph as described in Chapter 8. Keep a record of the locations.

134

2. Apply emulsion as described in this chapter. Expose, develop, and restain (with Giemsa, see Chapter 7).
3. Relocate the same cells and photograph.

SPECIAL PROCEDURES

Removal of Autoradiographic Film

1. Soak (overnight) in xylene and remove coverglasses with a scalpel.
2. Wash in xylene and absolute alcohol. Air dry.
3. Distilled H_2O, 10 min.
4. Potassium ferricyanide (7.5%), 3 to 4 min.
5. Sodium thiosulfate (20%), 5 min.
6. Two changes in distilled H_2O, 1 min each.
7. Trypsin (DIFCO, 1:250) 0.1% in phosphate buffer (pH 7.2 to 7.8) at 37°C, 10 min.
8. Wash in tap H_2O, 10 min.
9. Wash in distilled H_2O; dry.

Reference: Schmid, W.: Autoradiography of Human Chromosomes. In: *Human Chromosome Methodology,* J. J. Yunis, ed. New York, Academic Press, 1965, p. 107.

Removal of Silver Grains

Follow steps 1 through 6 for removal of autoradiograph film. This method is applicable to stripping film.

or

1. Clean slides as needed in xylene and absolute alcohol. Air dry.
2. Dip 15 sec in 0.75% $K_3Fe(CN)_6$ (potassium ferricyanide).
3. Place 3 min in 20% $Na_2S_2O_3$ (sodium thiosulfate).
4. Wash thoroughly in distilled H_2O, air dry completely, restain, and remount as needed.
5. An occasional grain remains after this method but does not interfere with photography or analysis of a metaphase after grain removal if the metaphase is restained. This method is applicable to liquid emulsion.

Method of: Gall, J. G., and Johnson, W. W.: Is There "Metabolic" DNA in the Mouse Seminal Vesicle? J. Biophys. Biochem. Cytol., 7:657, 1960.

135

Putting Emulsion on Slides Covered with Immersion Oil (or Mounted Slides)

1. Place in 2 changes of fresh xylene, about 2 min each. (Longer if coverslip is on slide.)
2. Rinse in absolute, 95%, 70%, 50% alcohol.
3. Rinse in running tap H_2O for at least 4 hr.
4. Rinse in distilled H_2O.
5. Air dry.
6. Apply emulsion.

Examination of Chromosomes* Obscured by Heavy Concentration of Silver Grains

1. Prepare the specimen for autoradiography as described in this chapter.
2. Cover each slide with a film of Vinalak, which is a synthetic resin dissolved in a volatile solvent immiscible with H_2O. Allow a small drop of the lacquer to fall on the surface of clean, distilled H_2O, where it spreads and forms a very thin film of resin.
3. Bring the slide, with cells uppermost, up under the resin film and withdraw the slide from the H_2O, carrying the resin film with it.
4. As the slide is lifted from the H_2O, tilt it at an angle of about $20°$ to the H_2O surface, so that one long edge emerges, allowing the H_2O to drain from between the resin film and the slide.
5. Dry at room temperature.
6. Apply, expose, and develop AR-10 stripping film as described in this chapter. (It is important to prevent the stripping film from floating off the slide during developing. Place a clean slide at the back of the specimen slide and a 1″ glass square at either end and in front of the specimen slide; hold together with 2 rubber bands.)
7. Mount dry slides (after removal of glasses and rubber bands) in Euparal.
8. After photographic record, float off the coverglass and stripping film by soaking in Euparal essence, leaving the resin film undamaged.

---

* This method may also be used for sex chromatin preparations.

9. The resin has a negligible effect on resolution, but if desired, it may be removed by taking the slide through Cellosolve to xylol, and leaving in xylol for 12 to 20 min.

10. Mount in Canada balsam.

Method of: Bishop, A., and Bishop, O. N.: Analysis of Tritium-Labeled Human Chromosome and Sex Chromatin. Nature, *199:*930, 1963.

10

Sex Chromatin

Sex chromatin is not well seen after hypotonic treatment. Therefore, chromosome preparations cannot be used to study sex chromatin. Chromosome stains are in general suitable for demonstrating sex chromatin but usage has caused some separation into sex chromatin and chromosome stains. Since it is a relatively dense nuclear structure, sex chromatin is visible with phase contrast microscopy in living cells. In fixed and stained preparations it is often, but not always, peripheral in location within the nucleus and stains darkly with nuclear stains. Each sex chromatin body represents an X chromosome; therefore sex chromatin examination is a useful procedure for the diagnosis of X chromosome abnormalities. Sex chromatin is visible only during interphase, but chromosomes are visible as dense structures only during metaphase. Thus many types of cellular tissues unsuitable for chromosome analysis are nevertheless good for sex chromatin evaluation. The interpretation of sex chromatin preparations is discussed in some detail under the section in this chapter on buccal smears. Modifications of the analysis for special tissues or preparations are also discussed under the appropriate sections.

138

References: Barr, M. L.: Sex Chromatin Techniques. In: *Human Chromo-some Methodology,* J. J. Yunis, ed. New York, Academic Press, 1965, p. 1.

Moore, K. L., ed: *The Sex Chromatin.* Philadelphia, W. B. Saunders Co., 1966.

FIXATION FOR SEX CHROMATIN EXAMINATION

More than one type of fixative is satisfactory for sex chromatin examination. The choice of fixative depends to some extent on the tissue specimen to be examined, whether smears or tissue blocks.

Reference: Culling, C. F. A.: Staining Affinities and Cytochemical Properties of the Sex Chromatin. In: *The Sex Chromatin,* K. L. Moore, ed. Philadelphia, W. B. Saunders Co., 1966, Ch. 5.

Modified Davidson's Fixative

Some investigators prefer Davidson's fixative to 10% buffered forma-lin for tissue blocks. It may also be used for smears.

FIXATION

Formalin (40% formaldehyde)	20%
95% ethanol	35%
Glacial acetic acid	10%
Distilled H_2O	35%

Tissue blocks should be fairly small, about $0.5 \times 0.5 \times 0.5$ cm, and should be placed in the fixative as soon as possible. Fix for 24 hr. Place in 70% alcohol for 12 hr to 3 days.

PROCESSING, EMBEDDING, AND SECTIONING

1.	95% alcohol	6 to 12 hr
2.	Absolute alcohol	1 to 2 hr
3.	Absolute alcohol	3 hr
4.	Absolute alcohol and xylene(equal parts)	1½ hr
5.	Xylene until clear	15 to 60 min
6.	Xylene and Paraplast (equal parts) in oven at 56°C	15 min
7.	Paraplast (1st change)	30 min
8.	Paraplast (2nd change)	30 min
9.	Paraplast (3rd change)	30 min
10.	Embed in Paraplast and cool quickly	
11.	Section at 5 μ; mount sections on clean, grease-free slides and dry overnight in an oven at 37°C.	

10% Buffered Formalin (for tissue blocks)

1. 100 ml formalin (formaldehyde 37% to 40%), 900 ml distilled H_2O, magnesium carbonate to excess to prepare 10% neutral formalin

2. 10% neutral formalin   1 liter
   Sodium dihydrogen phosphate, monohydrate   4 gm
   Disodium phosphate, anhydrous   6.5 gm

Fix 24 hr. Place in 70% alcohol for 12 hr to 3 days. Embed routinely or as described under modified Davidson's fixative.

95% Ethanol (for smears)

Fix at least 15 min.

140

Equal Parts 95% Ethanol and Ether (for smears)

Fix at least 15 min.

1 Part Glacial Acetic Acid: 3 Parts Absolute Methanol (acetic alcohol)
45% Acetic Acid

These two fixatives are used for chromosome preparation and are also satisfactory for fixation of tissue culture cells for sex chromatin analysis. Fix small buttons of cells or cells on coverglasses for about 15 min.

SEX CHROMATIN STAINS

Many stains are used for sex chromatin. Most of them are not specific for DNA but rely on the fact that relatively large amounts of chromatin (such as the sex chromatin body) will stain distinctively with a nuclear stain. Cytoplasmic counterstains are optional and may interfere with the analysis of sex chromatin. Some of the sex chromatin stains are described here.

References: Culling, C. F. A.: Staining Affinities and Cytochemical Properties of the Sex Chromatin. In: *The Sex Chromatin,* K. L.
Moore, ed. Philadelphia, W. B. Saunders Co., 1966, Ch. 5.

Gurr, E.: *Staining. Practical and Theoretical.* Baltimore, The
Williams and Wilkins Co., 1962.

Aceto-orcein and Lacto-aceto-orcein

(for smears and squash preparations)

The aceto-orcein and lacto-aceto-orcein method is described in the section on buccal smears—squash technique, this chapter. Aceto-orcein stain is also described in Chapter 7, under chromosome stains and under preparation of chromosomes by the squash technique.

Biebrich Scarlet-Fast Green (BS-FG)

(method of Guard for smears)

REAGENTS

1. Biebrich scarlet stain

Biebrich scarlet (water soluble, Harleco)	1 gm
Phosphotungstic acid	0.3 gm
Glacial acetic acid	5.0 ml
50% ethanol	100.0 ml

141

2. Fast green stain

Fast green FCF (Harleco)	0.5 gm
Phosphomolybdic acid	0.3 gm
Phosphotungstic acid	0.3 gm
Glacial acetic acid	5.0 ml
50% ethanol	100.0 ml

METHOD

1. From 95% ethanol fixation transfer to 70% ethanol for 2 min.
2. Stain in Biebrich scarlet for 2 min.
3. Rinse in 50% ethanol.
4. Differentiate in fast green FCF for 1 to 4 hr (usually 3 hr) until cytoplasm and nuclei are green (pyknotic nuclei and sex chromatin stain red).
5. Rinse in 50% ethanol and leave for 5 min.
6. Dehydrate in 70%, 95%, and absolute ethanol for 2 min each.
7. Clear in 3 changes of xylol for 2 min each.
8. Mount.

Carbolfuchsin Stain

(method of Eskelund, for sections and smears)

This technique is described in the section on newborn screening methods for examining sex chromatin, this chapter.

Cresylecht Violet

(Moore's method, for buccal smears)

1. Remove slides from 95% ethanol fixative and pass through 70% alcohol, 50% alcohol and distilled H_2O, 5 min in each, with 2 changes of distilled H_2O.
2. Immerse slides in a 1% solution of cresylecht violet (Coleman and Bell) for 5 to 8 min.
3. Differentiate in 95% alcohol, 5 to 8 quick dips.
4. Continue differentiation in absolute alcohol, checking with the microscope at intervals until the details of nuclear structure are defined clearly; usually this takes about 1 min.

142

5. Clear in 2 changes of xylol, 3 min in each.

6. Mount.

Feulgen Reaction

(for sections and smears)

This stain is the most specific for chromatin and sex chromatin. A slightly different modification of this technique is described for the staining of chromosomes (Chapter 7).

REAGENTS

1. Schiff reagent

    Dissolve 1 gm basic fuchsin in 400 ml distilled H_2O, using heat if necessary.

    Add 1 ml of thionyl chloride ($SOCl_2$), stopper the flask, shake, and allow to stand for 12 hr.

    Add 2 gm of activated charcoal; shake and filter immediately. This reagent keeps for several months in a well-stoppered, dark bottle in a refrigerator.

2. Sulfite rinses

    Sulfite rinses must be freshly prepared by adding 7.5 ml of 10% potassium metabisulfite and 7.5 ml of 1 N HCl to 135 ml distilled H_2O. This amount will fill 3 Coplin jars.

METHOD

1. Bring sections or smears to H_2O.

2. Rinse in cold 1 N HCl (optional).

3. Hydrolyze in 1 N HCl preheated to 60°C and leave for 10 min.

4. Rinse in cold 1 N HCl (optional) and rinse briefly in distilled H_2O.

5. Transfer to Schiff reagent for 30 to 90 min.

6. Place in sulfite rinse for 1 min.

7. Place in two more sulfite rinses, 2 min each.

8. Rinse well in distilled H_2O.

9. Counterstain, if desired, with 1% aqueous light green.

10. Dehydrate, clear, and mount.

143

Hematoxylin and Eosin

(for paraffin sections and smears)

REAGENTS

1. Harris's hematoxylin

Hematoxylin 10% in absolute alcohol	5.0	ml
Mercuric oxide	0.25	gm
Potash alum 10% aqueous	100.0	ml
Glacial acetic acid	4.0	ml

Mix well; allow to ripen in a 500-ml closed bottle for 3 months. then add the mercuric oxide, and when the solution turns deep purple, turn off the heat; then cool and add the acetic acid.

or

Ehrlich's hematoxylin

Hematoxylin 2% in absolute alcohol	100 ml
Potash alum 2.5% aqueous	100 ml
Glycerin	100 ml
Glacial acetic acid	10 ml

Mix well; allow to ripen in a 500 ml closed bottle for 3 months. Alternatively, dissolve 1 gm Ehrlich hematoxylin powder in a mixture of 4 ml glacial acetic acid, 80 ml 50% alcohol, and 40 ml glycerin; filter. The solution requires no ripening and is ready for immediate use.

2. Eosin, 1% aqueous solution.
3. Acid alcohol (1% HCl in 70% alcohol).
4. Base (1.5% sodium bicarbonate).

METHOD

1. Place thin paraffin sections (5 to 7μ) in xylol for 1 min, then in absolute alcohol for 30 sec, followed by a further 30 sec in 95% ethanol.
2. Wash paraffin sections and smears in distilled H_2O for a few seconds.
3. Stain in hematoxylin for 5 to 15 min (the longer time gives a deeper staining to the chromatin).
4. Rinse in H_2O.
5. Differentiate for a few seconds in acid alcohol.

144

6. Rinse in H_2O.
7. Blue in base (hematoxylin is red in acid solution).
8. Rinse in water and examine under a microscope to ensure that only the chromatin is stained. If the preparation is under-decolorized, repeat steps 5 to 8; if preparation is overdecolorized, steps 3 to 8 should be repeated.
9. Stain in eosin for 10 to 60 sec. The correct time will give a pale pink color to the cytoplasm, connective tissue, etc.
10. Rinse in H_2O.
11. Dehydrate rapidly in 95% and absolute alcohol.
12. Clear in xylol and mount in Harleco synthetic resin or DePex (DPX). A thin (No. 1 or 1½) coverslip should always be used to mount sex chromatin so that the oil immersion objective may be used.

Thionin Stain with Acid Hydrolysis

(method of Klinger and Ludwig, for sections and smears).

The technique for thionin stain with acid hydrolysis is described in the sections on buccal smears and sex chromatin examination of surgical and autopsy tissues, this chapter. The pH of this stain is critical.

THIONIN WORKING SOLUTION

1. Thionin stock solution: 1 gm thionin in 100 ml 50% ethanol.
2. Michaelis buffer solution: 9.7 gm sodium acetate $\times$ 3 H_2O, 14.7 gm sodium barbiturate in 500 ml of boiled distilled H_2O.
3. Bring 28 ml of buffer and 32 ml of 0.1 N HCl to 100 ml with thionin stock solution. pH should be 5.7 $\pm$ 0.2.

BUCCAL SMEARS

Buccal smears are recommended for routine sex chromatin examination.

Method

1. Microscope slides (precleaned in 95% ethanol) are marked ahead of time with code number or patient's name. Mark right (R) and left (L) if the 2 sides of the mouth are to be distinguished. Mark lip (lip) if this area is used in a child or un-

cooperative person. At least 2 slides should be prepared on each person. The slide identification also orients to the side with the specimen on it.

2. Use a metal spatula or, less preferably, the sharp side of a split wooden tongue depressor or a spoon. Scrape the buccal mucosa vertically by pressing *firmly* against fingers held on the outside of the cheek. The first scrapings may be discarded, or the patient may be asked to rinse his mouth out before the scrapings are taken.

3. Pool several scrapings on the slide and spread quickly as a thin layer over a large area of the slide. A second slide may be used to spread the scrapings, but this may tend to collect the cells unevenly. Mayer's egg albumin on the slide as an adhesive is unnecessary if a large number of cells are obtained.

4. Fix immediately and for at least 15 min in a Coplin or other jar containing 95% ethanol or equal parts 95% ethanol and ether. The slides of each patient should be fixed in a separate jar. If slides are to be transported, they may be removed from the fixative after 15 min and air dried, or they may remain in fixative up to 24 hr. Evaporation of the fixative should be avoided. Various methods are available to keep the cells from drying out on the slides if there is a long interval before staining. (See Culling, C. F. A.: In: *The Sex Chromatin*, K. L. Moore, ed. Philadelphia, W. B. Saunders Co., 1966, Ch. 5.)

5. Stain with any sex chromatin stain recommended for smears. A simplified version of the thionin stain with acid hydrolysis is described here. A positive control slide should be included.

6. Air-dried slides fixed for 15 min in 95% ethanol are hydrolyzed in 5 N HCl at room temperature for 20 min. Pass through 4 changes of fresh distilled H_2O. Handle carefully to prevent loss of cells from slide. pH of stain is critical; therefore all HCl must be rinsed off. Stain in thionin working solution (page 145) for 25 min. Rinse in distilled H_2O, 70% ethanol, 95% ethanol; blot dry on filter paper. If necessary, mount from xylene with No. 1 thickness coverglass.

Interpretation

1. Under low power objective, locate thin sheets of cells which are not superimposed or folded.

2. Switch to oil immersion objective and examine 100 to 200 cells. Avoid fields with few cells. Avoid folded, darkly stained, or poorly stained nuclei. Clumped chromatin in these cells can make distinction of the sex chromatin difficult.

Table 10-1. Some of the More Common Sex Chromatin Abnormalities in Buccal Smears

Sex Chromatin Body	Report	Prototype Chromosome Abnormality	Clinical Name
1 present in a male (15 to 35% of nuclei)	chromatin positive	XXY	Klinefelter syndrome
0 present in a female	chromatin negative	XO	Turner syndrome
1 present in a male (15% or less of the nuclei)	chromatin positive (note: that % is low)	XY/XX or XY/XXY	Genetic intersex or Mosaic Klinefelter syndrome
1 present in a female (15% or less of the nuclei)	chromatin positive (note: that % is low)	XO/XX or XY/XX	Mosaic Turner syndrome or Genetic intersex
2 present in a male	chromatin positive (note: double sex chromatin bodies)	XXXY	Klinefelter syndrome
2 present in a female	chromatin positive (note: double sex chromatin bodies)	XXX	Triple X female
1 present in a female (too large)	chromatin positive (note: too large)	XX (one X is an isochromosome)	Turner syndrome (usually)
1 present in a female (too small)	chromatin positive (note: too small)	XX (one X is deleted)	Turner syndrome (usually)

147

3. In buccal smears the sex chromatin body is defined as a staining or chromatin positive mass, about 1μ in diameter, on the inner surface of the nuclear membrane. Because nuclei are more difficult to read in buccal smears than in other types of preparations (such as amnion cells or tissue culture cells grown on coverslips), it is usually unwise to read nonperipheral sex chromatin in buccal smears. The color of the chromatin positive mass will depend on the stain used.

4. Routine reports should read: chromatin positive or chromatin negative. In most laboratories, at least 15% and not more than 30 to 35% of the readable nuclei in chromatin positive smears contain sex chromatin bodies as defined under point 3. In chromatin negative smears there are no nuclei containing sex chromatin bodies as defined under point 3. Any positive specimens that are 15% or less should be repeated. Any specimens that are discrepant with the phenotypic sex should be repeated, preferably 3 times.

5. Special situations include (a) chromatin positive specimens with percent of positive cells 15% or less, (b) abnormal number of sex chromatin bodies, and (c) abnormal shape of sex chromatin bodies (see Table 10-1). In these special situations record the number and percent of readable nuclei in the following categories:

0 Sex Chromatin Body		1 Sex Chromatin Body		2 Sex Chromatin Bodies		More Than 2 Sex Chromatin Bodies (Record number.)		Abnormal Size (Record large or small.)		Other (Specify.)	
#	%	#	%	#	%	#	%	#	%	#	%

Aceto-orcein Squash Technique for Buccal Smears

REAGENTS

1. Stock aceto-orcein solution
   Synthetic orcein   1 gm
   Glacial acetic acid   45 ml
   Boil, cool, and filter.

2. Working solution
   Dilute 45 parts stock solution with 55 parts distilled H_2O and filter. Periodic filtration is necessary.

3. Lacto-aceto-orcein working solution

> Dilute 50 parts stock solution with 50 parts 70% lactic acid and filter.

METHODS

1. Obtain oral mucosal smear by scraping the inside of the patient's cheek with a wooden or metal spatula. Make several short vertical strokes with firm pressure. Discard the first scrapings. The other hand may be used on the outside of the cheek to support firm pressure with the spatula.

2. Spread desquamated cells and saliva on the surface of a clean, dry microscope slide.

3. Quickly place one drop of aceto-orcein working solution on the center of the smear and a thin coverglass over it. (22 mm square or 22 × 40 mm rectangle, No. 1 thickness.)

4. Place 1 or more layers of filter paper on top and press the coverglass down by drawing the thumb across from one end of the slide to the other. To avoid moving the coverglass on the slide, hold the filter paper down firmly over one corner of the coverglass with the left hand, while the right moves across the slide. With correct pressure the nuclei are evenly flattened.

5. Store in covered tray in a refrigerator. Add fresh stain to the edge of the coverglass to prevent drying. For storage as long as 6 weeks at room temperature, use lacto-aceto-orcein working solution. The hydroscopic lactic acid prevents drying out.

6. Temporary preparations with either stain may be examined within 5 min. Simple aceto-orcein stains the nucleus and cytoplasm more rapidly than lacto-aceto-orcein. The staining deepens with time and examination is usually facilitated. However, contaminating bacteria, which may be the size of sex chromatin, also become more prominent with time.

7. Slides may be made permanent by inverting in a dish of 1:3 mixture of glacial acetic acid and absolute alcohol until the coverglass separates. Cells adhere to both slide and coverglass. Therefore, wash both 3 times for 2 min each in absolute alcohol, clear in xylol, and mount.

Method of: Sanderson, A. R., and Stewart, J. S. S.: Nuclear Sexing with Aceto-Orcein. Brit. Med. J., 2:1065, 1961.

MODIFICATIONS

1. Some investigators do not feel that pressure over the coverglass is necessary to flatten the cells. The pressure should be just heavy enough to express excess stain.
2. For permanent slides, the coverglass may be floated off with 45% acetic acid and the slide rinsed with drops of 45% acetic acid until dye does not appear to run off. Then air dry and mount from xylene (both slide and coverglass).
3. The coverglass of a temporary preparation may be ringed with Krönig cement (Fisher Scientific Co.).

SEX CHROMATIN EXAMINATION OF SURGICAL AND AUTOPSY TISSUES, INCLUDING ABORTIONS

The examination of sex chromatin directly in multiple tissues, either at autopsy or surgery, is important for the evaluation of sex chromosome abnormalities, including mosaicism. A method using modifications of the thionin stain of Klinger and Ludwig is described here.

Procedure

1. Fix tissues in 10% neutral buffered formalin and embed in paraffin. Section at 5 μ. (See this chapter, the section on fixation.)
2. Include a known sex chromatin positive control slide each time the staining procedure is performed.
3. Deparaffinize in clean xylene twice for 5 min each time.
4. Rinse in absolute ethanol, and dip in 0.2% paralodion in 1 part ether to 1 part 95% ethanol, air dry for 15 sec. Rehydrate in 70% ethanol and distilled H_2O.
5. Hydrolyze in 5 N HCl for 20 min at room temperature.
6. Rinse 4 times in distilled H_2O to be sure no acid is carried into thionin stain.
7. Stain in freshly prepared thionin working solution at pH 5.7 (page 145) for 10 min (or more).
8. Rinse quickly in distilled H_2O.
9. Air dry or dehydrate by passing through 50%, 70%, 95%, and absolute ethanol and into xylene.
10. Mount with synthetic resin.

150

Modifications

1. Tissues vary in the amount of stain removed by water and alcohols after staining. This variation can be monitored by gross or microscopic inspection of the slide. Heavily stained tissues may be left longer in 50% or 70% ethanol. Lightly stained tissues may be dehydrated in the alcohols more quickly or, when necessary, rehydrated through the alcohols and restained in thionin without repeating hydrolysis.

2. Eosin may be used as a counterstain for the identification of cellular and tissue morphology. After staining in thionin, rinse quickly in H_2O, and stain in eosin (0.5% eosin in 80% ethanol containing 0.004% glacial acetic acid) for ten seconds. Dehydrate quickly in 95% and absolute ethanol, clear in xylene, and mount.

Interpretation

1. Close attention must be given to the problem of what nuclei are readable and what nuclei are not readable. A readable nucleus is lightly stained, usually large, and relatively deficient in heterochromatin other than the sex chromatin. An unreadable nucleus is diffusely heterochromatic, small, and darkly stained.

2. The sex chromatin is read as positive only if it is well defined and adjacent to the nuclear membrane and averages about 1 μ in diameter. It may be closely applied to and drawn out along the nuclear membrane, or it may appear semicircular or triangular in shape with one edge "forming" the nuclear membrane. In elongated nuclei the sex chromatin is frequently present at the extreme tip of the nucleus.

3. The tissue is scanned under oil immersion magnification and read as positive or negative without a quantitative breakdown of percent of positive nuclei.

4. Special procedures, interpretation, and recording of data are indicated in the case of abortion sex chromatin studies. There is a great need for uniformity of reporting. The report of the Geneva Conference, "Standardization of Procedures for Chromosome Studies in Abortion" should be consulted. (See reference below.)

151

References: Geneva Conference. Standardization of Procedures for Chromosome Studies in Abortion. Cytogenetics, *5:*361, 1966.

Klinger, H. P., and Ludwig, K. S.: A Universal Stain for the
Sex Chromatin Body. Stain Techn., *32:*235, 1957.

Priest, J. H., Cooper, J. B., and Priest, R. E.: Autopsy Sex
Chromatin: The Diagnosis of Klinefelter's Syndrome. Arch.
Path., *81:*281, 1966.

NEWBORN SCREENING METHODS FOR EXAMINING SEX CHROMATIN

The buccal smear technique can be applied to newborns. However, the use of amniotic membranes has advantages in that (1) sex chromatin evaluation can be performed without disturbing the newborn baby; (2) placentas can be routinely obtained on all deliveries, including stillborns and newborns who die within hours after birth; (3) the percent of chromatin positive cells in amnions of normal female babies is about 90%, and thus the discrimination of normal chromatin negative males or chromatin positive mosaicism becomes a much easier task; and (4) repeat slides may be made by simply taking another piece of amnion from the placenta.

Procedure

1. Refrigerate placentas at the time of delivery. Collect once a day.
2. In the laboratory, remove a strip of amnion from the fetal side of the placenta. Eliminate as much as possible of the gelatinous material from the underside of the amnion.
3. Cut an 8 cm square and place in dish containing physiological saline.
4. After 20 min place amniotic membrane, gelatinous side up, on a glass plate and stretch over the surface. With a gloved finger, apply pressure to remove as much of the gelatinous material as possible.
5. Place a clean microscope slide face down on the membrane, and draw the sides of the membrane up around to the back of the slide. Place a second slide over the back of the first and fasten in place with a clip, to keep the membrane from slipping off the slide during fixation and staining.
6. Fix in 95% ethanol for 15 min.
7. Immerse slide in absolute ethanol for 3 min and in 0.2% paralodion (dissolved in equal parts by weight of absolute alcohol and ether) for 2 min.

8. Air dry for 15 sec, place in 70% ethanol for 5 min, and 2 rinses of distilled H_2O for 5 min each.
9. Hydrolyze in 5 N HCl for 15 min.
10. Rinse 2 times in distilled H_2O, 2 min each.
11. Stain for 9 min in carbolfuchsin.
12. Place in 95% ethanol 1 min, absolute ethanol 1 min, xylene 30 min, remove and mount with Permount.

Method of: Robinson, A., and Puck, T. T.: Studies on Chromosomal Non-disjunction in Man. II. Amer. J. Hum. Genet., *19*:112, 1967.

Carbolfuchsin Stock Solution

3% fuchsin in 70% alcohol	10 ml
5% phenol	90 ml

Carbolfuchsin Working Solution

Stock solution	45 ml
Glacial acetic acid	6 ml
35% formaldehyde	6 ml

Mix and allow to stand for 24 hr before use. This solution keeps several months.

SEX CHROMATIN EXAMINATION OF CELLS FROM THE REPRODUCTIVE TRACT

Vaginal smears can be used for the examination of sex chromatin, using modifications of the well-established Papanicolaou techniques. Only a nuclear stain is necessary for the study of sex chromatin. At the time of gynecological examination, sex chromatin evaluation is especially indicated in cases of primary amenorrhea and malformations of the external genitalia.

1. Sex chromatin is most conspicuous in large vesicular nuclei of the basal layer of cells from the vaginal wall. Superficial cells contain dark, pyknotic nuclei which cannot be analyzed for sex chromatin. Therefore material should be collected from the deepest possible cell layers. This collection can be made by using a metal spatula with a rather sharp edge. The cells should be scraped from the vaginal wall with some force, and a "surface biopsy" obtained.

2. Smear on a clean slide with a few gentle strokes. Slides coated with Mayer's albumin may be used. (See section, this chapter on buccal smears.)

3. Fix immediately for 15 min in equal parts of 95% ethanol and ether, or 95% ethanol. (See pages 139-141 on fixation.) Slides may remain in fixative for at least 24 hr or may be air dried before staining.

4. Stain with sex chromatin stain for smears. (See section, this chapter on stains.)

Method of: Carpentier, P. J.: Sex Chromatin in Smears from the Reproductive and Urinary Tracts. In: *The Sex Chromatin,* K. L. Moore, ed. Philadelphia, W. B. Saunders Co., 1966, Ch. 10.

SEX CHROMATIN EXAMINATION OF CELLS FROM URINE

1. Voided urine may be kept at room temperature for about 6 hr without alteration of cellular morphology. If a longer time is anticipated before processing, add 30% 1-propanol solution, 3 parts to 1 part of urine.

2. Aspirate urine into a 20-ml hypodermic syringe and push through a Millipore filter (type SM, 13 mm diameter, pore size 5.0 $\pm$ 1.2μ) in a Swinny hypodermic adapter (Millipore Filter Corp.).

3. When a slight resistance is felt, remove the filter, secure to a glass slide with a metal clip, and fix immediately in 95% ethanol for 30 min.

4. Stain immediately or leave in fixative for several days.

Method of: Collett-Solberg, P. R., and Grumbach, M. M.: A Simplified Procedure for Evaluating Estrogenic Effects and the Sex Chromatin Pattern in Exfoliated Cells in Urine: Studies in Premature Thelarche and Gynecomastia of Adolescence. J. Pediat., *66:*883, 1965.

DRUMSTICKS IN POLYMORPHONUCLEAR LEUKOCYTES (NEUTROPHIL SEX CHROMATIN TEST)

The term *sex chromatin* comprises two superficially dissimilar structures: the Barr body or sex chromatin body of interphase nuclei, present in epithelial and other tissue cells; and the drumstick of the polymorphonuclear leukocytes. A drumstick consists of a small nuclear mass, about 1.5 μ in diameter, which is attached to the body of the nucleus by means of a thin filament. It is visible in peripheral blood smears prepared routinely. The analysis of drumsticks presents certain difficulties for screening X chromosome abnormalities. The number of cells containing drumsticks is variable in the normal female, and the relation between the number of drumsticks and the number of X chromosomes in excess of one is not always direct. Drumstick analysis may be useful in the evaluation of mosaicism for X chromosome abnormalities because a different tissue can be "sexed" by this method.

References: Davidson, W. M.: Sexual Dimorphism in Nuclei of Polymorphonuclear Leukocytes in Various Animals. In: *The Sex Chromatin*, K. L. Moore, ed. Philadelphia, W. B. Saunders Co., 1966, Ch. 3.

Mittwoch, U.: Sex Chromatin. J. Med. Genet., *1:*50, 1964.

INTRAUTERINE DIAGNOSIS OF SEX BY SEX CHROMATIN EXAMINATION

Cells in amniotic fluid may be used to determine the sex of the fetus. In X-linked disorders, therapeutic abortion may be indicated if the fetus is male.

1. Divide 10 ml of amniotic fluid between 2 tubes and centrifuge. Process as soon as possible after the fluid is obtained, preferably within 1 to 2 hr.

2. Transfer the sediment to protein-coated slides. Using caution to avoid washing away material, place the slides for 15 min in fixative (equal parts of 95% ethanol and ether).

3. Stain with sex chromatin stain for smears. (See section, this chapter, on stains.)

Method of: Riis, P., and Fuchs, F.: Sex Chromatin and Antenatal Sex Diagnosis. In: *The Sex Chromatin*, K. L. Moore, ed. Philadelphia, W. B. Saunders Co., 1966, Ch. 13.

SEX CHROMATIN IN PRIMARY OUTGROWTH FROM EXPLANTED TISSUE CULTURE FRAGMENTS

If a coverglass is placed firmly over a primary explant, the cells that grow out will attach to the coverglass. Methods for preparing coverglass cell preparations are described in Chapter 6, the section on short-term cultures from small tissue biopsies, coverglass method. These primary cell outgrowths may have overlapping cells, and the incidence of sex

155

chromatin in the cells tends to show greater variation than is true when cells in serial culture are grown on coverglasses.

Reference: Miles, C. P.: The Sex Chromatin in Cultured Cells. In: *The Sex Chromatin*, K. L. Moore, ed. Philadelphia, W. B. Saunders Co., 1966, Ch. 12.

SEX CHROMATIN IN CELLS FROM LONG-TERM TISSUE CULTURES—COVERGLASS METHOD

The cells are very well spread by the coverglass method. Both peripheral and nonperipheral sex chromatin can be read in normal human diploid female fibroblast-like cells. The percent of positive cells is 90 ± 5.

1. Inoculate 3×10^5 cells per 60×15 mm sterile tissue culture plastic Petri dish, containing a sterile 22×40 mm coverglass precleaned in 95% ethanol. The total number of dishes prepared depends on the individual situation. For routine evaluation of sex chromatin, 4 to 6 coverslips are ample. (See Chapter 15 for details of the replicate plating technique.) In this situation the replicate nature of the plating from one dish to another is not critical.

2. Place the dishes in a humidified desiccator, gas as needed with CO_2 and incubate at 37°C until the cells are almost at saturation density (confluent over the coverglass). Move the dishes as little as possible during this time, especially when the cells are attaching to the coverglass during the first 24 hr.

3. Pour medium off coverglass in Petri dish. Add 2 to 3 ml balanced salt solution (PBS, see Chapter 12), swirl, pour off, add 2 to 3 ml PBS, swirl, pour off.

4. Fix in 95% ethanol for 10 to 15 min in the same Petri dish.

5. Remove coverglass from dish and air dry.

6. Stain coverglass with sex chromatin stain, or mount cell side up on a microscope slide and allow to dry well before staining. (See section on stains, this chapter.)

Method of: Priest, J. H.: unpublished.

SEX CHROMATIN IN CELLS FROM LONG-TERM TISSUE CULTURES—TRYPSINIZATION METHOD

The cells are less well spread by the method described in this section. Therefore, the percent of positive cells from the normal human diploid female is more variable and depends on how strictly a "readable cell" is defined. Usually the percent positive is at least 30.

1. Take a confluent bottle of monolayer cells, decant medium, rinse 2 times with balanced salt solution (PBS, see Chapter 12), add 1 ml 0.25% trypsin-0.25% Versene (see Chapter 12), incubate at 37°C until cells are dispersed.

2. Add sufficient PBS to suspend cells. Centrifuge 600 to 800 rpm for 8 min.

3. Remove all supernatant. Add about 2 ml of 1 part glacial acetic acid to 3 parts absolute methanol, without disturbing the button of cells. Fix 30 min.

4. Suspend cells by pushing out with Pasteur pipette. Centrifuge as before and remove all supernatant fluid.

5. Add 0.1 to 0.2 ml 45% acetic acid to the cell button. (Amount should be adjusted to make a moderately heavy cell suspension.)

6. Prepare air-dried slides by allowing a drop of cell suspension to run down a tilted slide, precleaned in 95% ethanol. Air dry quickly. More than one drop of cell suspension may be placed side by side on one slide.

7. Stain with sex chromatin stain. (See section on stains, this chapter.)

Method of: Priest, J. H.: unpublished.

11

Meiotic Chromosomes

Studies of human meiotic chromosomes present certain difficulties. Testicular and ovarian tissues are hard to obtain, and they may be difficult to process. Meiotic *metaphase I* chromosomes are more condensed than are mitotic metaphase chromosomes; therefore, morphology during meiotic metaphase has not been characterized in the same detail. Nevertheless, evaluation of human chromosomes in germ cells is important because at this stage abnormalities of pairing can be demonstrated. In addition, germ cell chromosome abnormality is direct evidence for the possibility of transmission to subsequent generations.

Reference: Ohno, S.: Direct Handling of Germ Cells. In: *Human Chromosome Methodology,* J. J. Yunis, ed. New York, Academic Press, 1965.

PREPARATION OF CHROMOSOMES FROM TESTICULAR MATERIAL

Squash Method of Ohno for the Study of 1st Meiotic Prophase Figures

First meiotic prophase figures can be found in the testis of the sexually mature male. Ohno recommends that these stages be examined

158

without hypotonic pretreatment. (See reference at the beginning of this chapter.)

PROCEDURE

1. Cut tissue into pieces no larger than 2 × 2 × 2 mm and suspend in isotonic solution at 3°C.
2. After 10 min transfer the pieces to fresh cold solution. Repeat 2 or more times.
3. Prepare fresh fixative by mixing equal volumes of glacial acetic acid and distilled H_2O. Fixative volume should be at least 50 times the volume of the tissue to be fixed.
4. Immerse tissue in fixative for 15 to 45 min.
5. Place one cube of fixed tissue on a glass slide along with about 0.1 ml of fixative. To release most of the free cells, tap the cube gently with a blunt metal instrument. Remove the stringy connective tissue remaining in the free cell suspension with a forceps.
6. Cover with a No. 1 coverglass (24 × 40 mm). Take care to avoid formation of air bubbles.

159

7. Place the covered slide in the center of 4 layers of No. 1 filter paper folded in the middle, on a completely flat surface.

8. Bring the full weight of the body evenly upon the two thumbs placed over the coverglass. Apply pressure straight downward for at least 1 min, preferably 2 min. Avoid any sidewise movement of the coverglass.

9. Immerse the slide for 1 min in a mixture of dry ice and methanol in a large beaker. The coverglass can then be pried off the frozen fixative with a razor blade and discarded. (Neither coating the coverglass with silicon nor the slide with albumin is necessary to keep well-squashed cells on the slide. Simply rub a clean slide vigorously with a coarse paper towel several minutes before use.)

10. Air dry the slide and immerse in methanol for 15 min to extract the fatty substances usually found in gonadal tissue.

11. Air dry again, wash in tap H_2O, and hydrolyze in 1 N HCl at 60°C for 15 min.

12. Stain for 3 hr in leuco-basic fuchsin or 5 min in Giemsa solution. (See Chapter 7, chromosome stains.)

13. Air dry and mount with No. 1 coverglass and synthetic balsam.

INTERPRETATION

At pachytene the normal XY-bivalent is heavily condensed and contrasts sharply with the 22 autosomal bivalents which demonstrate fine chromomeric patterns (Fig. 1-8, page 11).

Squash Method of Ohno for the Study of 1st and 2nd Meiotic Metaphase

Hypotonic treatment improves the quality of meiotic metaphases from the sexually mature male. (See reference on page 158.)

PROCEDURE

1. Immerse a block of fresh testicular tissue in isotonic solution. Under the dissecting microscope, remove individual seminiferous tubules at least 2 cm long.

2. Place one long piece of tubule in the middle of a glass slide, cover with 2 ml of double-distilled H_2O, and allow to set for 30 min.

3. At the end of this time the tubule is swollen to 3 times its original size. Absorb distilled H_2O with filter paper.

160

4. Place 0.5 ml 50% acetic acid on the tubule and fix 15 min.
5. Tease the tubule into small fragments with fine forceps.
6. Squash and stain as described under squash method of Ohno for 1st meiotic prophase figures in testicular tissue (this chapter).

INTERPRETATION

1. The X and Y are associated end-to-end during 1st meiotic metaphase of a spermatocyte (Fig. 1-9, page 12). There are 22 autosomal bivalents, each showing chiasmata.
2. Each 2nd meiotic metaphase contains 22 autosomes and either the X or Y. The sex chromosome is usually heavily condensed. The sister chromatids of autosomes are completely split apart except at the centromeric regions.

Flame-dry Method

1. Obtain tissue by biopsy or orchiectomy. Fresh autopsy tissue may also be suitable.
2. Place seminiferous tubules, immediately after removal from the patient, in a small Petri dish without adding any saline, and mince as finely as possible with sharp scissors.
3. Add about 2 ml of 0.6% hypotonic sodium citrate solution to tissue shreds with volume of about 0.5 ml or less.
4. Mix gently with a pipette and allow to stand at room temperature for 30 min.
5. Suspend cells in the hypotonic solution and fix by adding 6 ml of acetic alcohol (1:3) for about 30 min.
6. Centrifuge at 1000 rpm for 5 min, decant supernatant, and add 6 ml of fresh fixative.
7. Repeat fixative change at least twice at about 20-min intervals.
8. After final centrifugation resuspend the cell pellet in 2 ml fresh fixative, mix well by pipetting, and then allow the larger tissue fragments to settle to the bottom of the tube.
9. Remove the upper layer of dispersed cell suspension and place 1 or 2 drops on a clean slide.
10. Hold the slide in a horizontal position and bring the edge of the slide into contact with a small flame of a Bunsen burner for several moments. Allow fixative to burn off in order to dry the slide completely.

161

11. Stain with Giemsa stain or other chromosome stain, and mount. (See Chapter 7.)
12. Prepare multiple slides (20 or more may be necessary).
13. Prepare at least 10 slides without applying the hypotonic treatment (for the study of 1st meiotic prophases).

Method of: Sasaki, M., and Makino, S.: The Meiotic Chromosomes of Man. Chromosoma, *16:*637, 1965.

Air-drying Method

The air-drying method is recommended for the examination of metaphase figures. There may be less cell breakage than with squash techniques.

1. Remove testicular material and place in isotonic sodium citrate solution (2.2% wt/vol) at room temperature. If necessary, pierce the tunica and swirl the testis in the solution to remove adherent fat.
2. Transfer testicular material to fresh 2.2% sodium citrate in a small Petri dish and gently pull out tubules. Hold the mass of tubules with fine, straight forceps and thoroughly tease out their contents with fine, curved forceps. When the tubules appear "flat" and opaque, allow them to settle; then transfer the supernatant fluid to a 15-ml centrifuge tube.
3. Centrifuge at 500 rpm for 5 min (radius of centrifuge head 7 in.).
4. Discard the supernatant fluid and resuspend the sedimented cells in about 3 ml of 1% sodium citrate solution.
5. Leave the cells in hypotonic solution for 12 min at room temperature.
6. Divide the suspension into 2 equal parts and transfer to 2-ml centrifuge tubes. Centrifuge at 500 rpm for 5 min, with slow acceleration.
7. Remove as much of the supernatant fluid as possible. Resuspend the cells in the remainder by flicking the tube with a finger, so that a thin film of suspension adheres to the wall of the tube.
8. Add about 0.25 ml fixative (1 part glacial acetic acid to 3 parts absolute ethanol, plus about 1 part chloroform to 40 parts of the acetic-alcohol mixture).
9. Again flick the tube to insure mixing. Continue to add fixative until the tube is about one-third full.

162

10. After 5 min, centrifuge and resuspend cells in fresh fixative. Repeat the change of fixative after about 10 more minutes.

11. Take up some of the suspension in a small pipette. Allow a drop to fall on a grease-free slide at room temperature. If the slide is thoroughly clean and the fixation satisfactory, the drop will expand evenly, reach a maximum, and then begin to retract. When interference colors are seen, blow gently on the slide to hasten the final evaporation.

12. Repeat with further drops. Several successive drops of a dilute suspension may give better preparations than single drops of more concentrated suspensions.

13. Stain in lactic-acid-orcein or other chromosome stain. (See Chapter 7.)

Method of: Evans, E. P., Breckon, G., and Ford, C. E.: An Air-Drying Method for Meiotic Preparations from Mammalian Testes. Cytogenetics, *3:*289, 1964.

Alternative Squash Method

The alternative squash method is suitable for the examination of pachytene stages.

1. Place a tissue sample in 0.3% sodium citrate solution for 1 to 6 hr, depending on the size of the specimen.

2. Stain, squash, and observe a few tubules at this stage to be sure spermatogenesis is present in the sample.

3. Place in a 3M solution of glucono-delta-lactone for 2 hr (to soften the tissue).

4. Stain with aceto-carmine or propiono-carmine 10 to 12 hr.

5. Wash out excess stain with 4 changes of 70% alcohol. Material may be stored at 0° to 5°C in the last wash of 70% alcohol for several months if necessary.

6. Place stained tissue in a shallow dish and cover it with equal parts mixture of absolute alcohol and glacial acetic acid. Mince it into a thick suspension.

7. Filter the cell suspension through several layers of cheesecloth.

8. Centrifuge the filtrate at 250 rpm for 15 min. Discard the upper half of the supernatant. With a pipette, draw off the fluid just above the coarse precipitate. This layer of fluid contains most of the spermatocytes and is used for immediate squashing or refrigerated storage.

163

9. Pipette 1 drop of cell suspension onto a siliconized slide.
10. Add a drop of Hoyer's medium and mix. (A water-soluble mounting medium, described in *Laboratory Manual for Introductory Mycology,* C. J. Alexopoulos and E. S. Beneke, eds., Minneapolis, Burgess Publishing Co., 1962.)
11. Apply a coverglass and blot gently with bibulous paper to remove excess fluid.
12. Heat gently on warming plate to flatten the cells and fix them to the slide.
13. Invert the slide on several sheets of bibulous paper and apply firm pressure with the thumb.
14. The slide may be rewarmed and the squash procedure applied again if necessary.
15. Squashing and permanent mounting are accomplished in a single step. Slides keep for at least 8 months.

Method of: Gardner, H. H., and Punnett, H. H.: An Improved Squash Technique for Human Male Meiotic Chromosomes: Softening and Concentration of Cells; Mounting in Hoyer's Medium. Stain Techn., *39:*245, 1964.

PREPARATION OF CHROMOSOMES FROM OVARIAN MATERIAL

Fetal Ovarian Tissue—Squash Method of Ohno for the Study of 1st Meiotic Prophase Figures

(See reference on page 158.)

PROCEDURE

1. Immerse fetal ovary in isotonic solution.
2. Under the dissecting microscope, remove the cortical area and cut into pieces no larger than 2 mm square.
3. Place the tissue pieces in a 10% trypsin solution and incubate 30 min at 37°C.
4. Rinse 3 times in isotonic solution, 10 min each.
5. Fix, squash, and stain as described under squash method of Ohno for 1st meiotic prophase figures in testicular tissue (this chapter).

INTERPRETATION

1. The normal female pachytene nucleus contains 23 bivalents, each demonstrating an intricate chromomeric pattern (Fig. 1-10, page 13).

164

2. Unlike the XY-bivalent, the XX-bivalent in fetal oöcytes does not condense.

3. There is a tendency in female nuclei for autosomal bivalents to form nonhomologous associations at their centromeric regions.

Oöcyte Preparation from a Mature Follicle—Squash Method of Ohno for the Study of 1st Meiotic Metaphase

(See reference on page 58.)

1. Fresh material from the human female is very difficult to obtain. This procedure has been successful with smaller mammals.

2. Puncture the mature follicle with the sharp points of a watch-maker's forceps. Pour out liquor folliculi into a Petri dish.

3. Under the dissecting microscope, locate an ovum surrounded by the cells of corona radiata.

4. Incubate the Petri dish at 37°C until most of the corona radiata cells detach from the ovum (about 50 min).

5. Under the dissecting microscope, suck out the ovum with a small pipette and deposit it with a small amount of fluid into a drop of 50% acetic acid on a glass slide.

6. Fix for 15 min and cover the slide with a coverglass. Only gentle squashing is needed to spread apart the bivalents.

Short-term Incubation of Ova Released from Their Follicles

References: Edwards, R. G.: Maturation *in Vitro* of Human Ovarian Oöcytes. Lancet, *ii*: 926, 1965.
Yuncken, C.: Meiosis in the Human Female. Cytogenetics, *7:* 234, 1968.

12

Principles of Tissue Culture

Some of the basic information needed for cell culture in a cyto-genetics laboratory is discussed in this chapter. Even *short-term culture* of human peripheral blood requires an investment in culture equipment and supplies and a knowledge of culturing principles.

References: Merchant, D. J., Kahn, R. H., and Murphy, W. H., Jr.: *Handbook of Cell and Organ Culture.* Minneapolis, Burgess Publishing Co., 1964.

Paul, J.: *Cell and Tissue Culture.* 3rd ed. Baltimore, The Williams and Wilkins Co., 1965.

Priest, J. H.: Human Cell Culture: An Important Tool for the Diagnosis and Understanding of Disease. J. Pediat., *72:*415, 1968.

White, P. R.: *The Cultivation of Animal and Plant Cells.* 2nd ed. New York, The Ronald Press Co., 1963.

Willmer, E. N., ed.: *Cells and Tissues in Culture.* Vols. I-III. New York, Academic Press, 1966.

HEAT STERILIZATION

Dry Air

1. 165°C for 2 hr.
2. Items to be sterilized:
   Tissue culture pipettes, graduated to tip, constricted to hold cotton plug.

Pasteur pipettes—disposable.
Instruments.

3. Dry air tape is available with markings that turn color with dry air sterilization.
4. Date of sterilization should be marked on the tape.

Autoclaving—Fast Exhaust

1. 250°F or 121°C for 20 to 30 min, 15# pressure.
2. Items to be sterilized:
   Empty glass containers,* caps loosened, top covered with aluminum foil.
   Desiccators—partially open.
   Small instrument packs.
   Rubber bulbs; other rubber items.
   Teflon cloning rings; other materials damaged by temperatures over 250°F.

Autoclaving—Slow Exhaust

1. 250°F or 121°C for 20 min, 15# pressure.

---

* Because prescription bottles are soft glass, they may break on fast exhaust, in which case they should be sterilized by slow exhaust.

2. Items to be sterilized:
   Bottles containing liquid, not more than 2/3 full, caps loosened.
3. Autoclaving tape is available with markings which turn color with autoclaving.
4. Date of sterilization should be marked on tape.
5. Millipore filter assemblies should always be autoclaved on slow exhaust, small filters for 15 min and large for 35 min. (See special instructions which come with the filters.)

General

1. Don't dry air sterilize and dry glassware in the same oven at the same time.
2. Assemble glassware, etc., in one area; place sterilized items in a different tissue culture area to cool, and put away as soon as possible.
3. Caps should be loose on autoclaved glassware until the containers are dry. Then be sure to tighten for storage.
4. Sterile supplies should be stored in specific areas (shelf, cabinet, etc.) marked "sterile." Nonsterile items should never, under any circumstances, be stored in the sterile areas.
5. As a general rule sterile supplies should be resterilized at intervals not to exceed one month. If they are in frequent use, such as tissue culture pipettes, they will be sterilized more frequently.
6. Aluminum foil is useful for covering items prior to dry or moist air sterilization.
7. A separate recording thermometer should be placed in oven or autoclave, to record the maximal temperature.

FILTER STERILIZATION

Small Volumes (less than 100 ml)

Filter holders of various sizes are available (Millipore Corp.; Swinny filter holder and Micro-syringe holder) to fit on a hypodermic syringe, for pressure filtration of small volumes of liquid. These filter holders are very useful; they should be assembled and used according to the accompanying instructions.

Intermediate Volumes (100 ml to 1000 ml)

Filter holders are available for suction filtration (Millipore Corp.; Pyrex filter funnel and base with coarse-grade fritted support for filter).

A vacuum pump may be necessary to furnish enough suction for rapid filtration unless the laboratory is equipped with adequate wall suction. These filter holders should be assembled and used according to the accompanying instructions.

Larger Volumes (over 1000 ml)

Large filter holders, which operate by pressure filtration, are available (Millipore Corp.) to hold 90 mm diameter filters, 142 mm diameter, and larger. The 142 mm diameter filter sterilizes up to 10 liters at a time and is useful for media filtration in a tissue culture laboratory of moderate size. A pressure holding vessel and a receiving vessel are needed, in addition to the filter holder, filter, and connections. Filters and accessory parts should be assembled and used according to the accompanying instructions.

General

1. Filters of 0.20 μ or 0.22 μ pore size should be used for sterilization (Millipore or Gellman catalogues). The rate of filtration is less than with filters with larger pores, but small Gram-negative bacteria are excluded.
2. Prefilters are needed if particulate material or serum is present.
3. Filter holders and accessory parts should be washed and rinsed immediately according to accompanying instructions. In addition, the general washing rules for tissue culture glassware apply to filters (see section, this chapter, on washing). Allow to air dry. Assembly (including positioning of the filter and prefilter) should take place as soon as the parts are dry to prevent accumulation of dust, etc.
4. Sterilization of larger filter assemblies should take place just before use, allowing enough time for cooling.
5. All filter assemblies require autoclaving with slow exhaust to keep the filter from breaking. The total autoclaving time depends on the size of the filter.

WASHING TISSUE CULTURE GLASSWARE

Protocols for the washing of tissue culture glassware vary considerably in complexity. A great deal depends on the individual laboratory and whether or not special needs are present. It is wise to buy disposable supplies that come sterile and may be discarded after use. The budget of the laboratory can often be adjusted to include disposable supplies rather than salary for a full-time dishwasher. There still remain some permanent items which come directly or indirectly into contact with tissue culture cells. The more complex protocols will not be included here. A compromise plan which has stood the test of time well will be presented.

General Procedure

1. It is important to soak the items immediately after use. Use 1% 7X solution (Linbro Chemical Co., New Haven, Conn.).
2. Scrub with fresh 1% 7X and remove all labels. Scrub off marking pencil with scouring pad.

   *

3. Boil 1 hr in fresh 7X (unless this is damaging to the item).
4. When cool, scrub again.
5. Rinse at least 1 hr in running tap H_2O.
6. Rinse a total of 3 separate times in single distilled H_2O.
7. Rinse a total of 3 separate times in triple distilled H_2O.
8. Air dry.
9. When dry, store covered or assemble for sterilization.

Washing Pipettes

1. Soak immediately in 1% 7X. Wash at least 2 times a week.
2. Follow instructions for automatic pipette washer.† The instructions here are for VirTis Automatic Pipette Washer (The Virtis Co., Inc., Gardiner, N. Y.).
3. Add 20 liters of hot tap H_2O and 200 ml 7X to tank.
4. Turn on heater unit to 180°F.

---

* Tissue culture glassware should be used only for tissue culture. New glassware, glassware exposed to cleaning solutions or strong reagents, and in some cases Petri dish bottoms, which are directly in contact with cells, should have other washing steps inserted at this point. Rinse 3 times with tap H_2O, boil 1 hr in 1% solution of EDTA (ethylenediaminetetraacetic acid, disodium salt), rinse 3 times in tap H_2O. EDTA is a noncolloidal organic chelating or complexing agent with the ability to deionize heavy metals and alkaline earth ions.

† A flush-type pipette washer attached to the hot H_2O tap may be used. It is important for the H_2O to be as warm as possible since 7X cleans most effectively at near-boiling or boiling temperatures. Alconox may be used, in place of 7X.

5. Place pipettes in rack with tips up. Place rack in washer chamber and start pump. Wash 30 min (cycling time 5 to 6 min).
6. Transfer rack to flush-type pipette rinser attached to a tap H_2O faucet and rinse for 1 hr.
7. Put rack in fresh single distilled H_2O, lift up, and drain entirely. Repeat to a total of 3 times.
8. Put rack in fresh triple distilled H_2O, lift up, and drain entirely. Repeat to a total of 3 times.
9. Transfer to oven or pipette dryer to dry.

DISTILLED WATER

The exact preparation of distilled H_2O for tissue culture varies from one laboratory to another. Certain rules must be followed, however, or cells will not grow, particularly mammalian diploid cells. Deionized water is poor for tissue culture use, as is single distilled H_2O from a standard Barnsted still. H_2O should be triple distilled in glass or quartz. A very satisfactory compromise procedure is to obtain single distilled H_2O from a Barnsted still and run through a double quartz still. (A German still, Heraeus Bi-Distiller, is available through Brinkmann Instruments, Westbury, N. Y. This double quartz still is available in two sizes which deliver 400 or 1500 ml per hr.)

TEMPERATURE, pH, AND HUMIDITY CONTROL

Mammalian cells require constant temperature conditions of 37°C to 37.5°C, pH of 7.2 to 7.4, and humidity close to saturation. Constant temperature is maintained in a water-jacketed incubator, provided the doors are opened infrequently. pH is usually maintained by a constant CO_2 atmosphere and bicarbonate buffer in the culture medium. A solution of phenol red in the medium is a sensitive indicator of pH. (The color of this indicator is pink-orange in the proper pH range. Yellow is too acid; pink is too alkaline.) A humidified gas flow incubator may be purchased, or flow gauges may be installed on air line and CO_2 tank to control the entrance of a constant CO_2-air mixture. This mixture is bubbled through H_2O in a pan inside the incubator, for humidification. Alternatively, tissue culture flasks and bottles or desiccators may be gassed with the appropriate concentration of CO_2 and closed tightly. Gas-tight tissue culture flasks and bottles maintain the proper humidity. If a desiccator is used, H_2O is placed in the bottom, and the lid is lubricated well with silicone grease. Desiccators are useful to hold tissue culture Petri dishes which tend to dry out and change temperature and pH quickly. The method of using gas-tight containers avoids loss of pH and humidity each time the incubator is opened and is less subject to technical failure. However, it is less efficient when a large number of cultures are processed.

MEDIA

Defined media for the culture of mammalian diploid cells are available commercially, or standard recipes may be used for assembling the ingredients. It is always important to use triple distilled H_2O. (See section this chapter on distilled H_2O.) Some of the larger companies that supply tissue culture media are: Difco Lab., Detroit, Mich.; Hyland Laboratories, Los Angeles, Calif.; Flow Laboratories, Rockville, Md.; Microbiological Associates, Bethesda, Md.; Grand Island Biological Co., Grand Island, N. Y., and BioQuest, Cockeysville, Md. Protocols for the various types of media are listed in the catalogues of these companies or in the tissue culture textbooks listed on page 166. The recipe for one defined medium will be given here (the medium of Dulbecco and Vogt). The growth of mammalian cells requires, in addition to the defined medium, a concentration of approximately 15% serum. Fetal calf serum (FC) is probably the best for human cells. Sera are also available from the same companies supplying defined medium.

Media Recommended for Human Cells

This list of media available commercially is not complete.

> Dulbecco and Vogt's modification of Eagle's medium (D & V)
> Medium 199
> NCTC-109
> Puck's F-10 or F-12
> Waymouth's medium

COMPOSITION AND PREPARATION OF DULBECCO AND VOGT'S MODIFICATION OF EAGLE'S MEDIUM (D & V). (Tables 12-1 and 12-2.) This recipe for Dulbecco and Vogt's modification of Eagle's medium is listed because it is one of the least complex to prepare of all the defined media for mammalian cells. The medium supports growth and differentiated function of many types of human cells. Single cell plating efficiency may be lower in this medium than in, say, F-12. (Single

172

cell plating efficiency is also a function of the fetal calf serum.) Sterilization of this medium by positive pressure filtration with 100% CO_2 produces a pH on the acid side (yellow color of phenol red). Some of

Table 12-1. Stock Solutions for Dulbecco and Vogt's Medium

Amino Acid Stock Solution for 50 Liters

	Grams	Final Conc. in Medium
l-arginine HCl	4.20	0.0840 gm/L
l-histidine HCl·H$_2$O	2.10	0.0420
l-isoleucine	5.24	0.1048
l-leucine	5.24	0.1048
l-lysine HCl	7.31	0.1462
l-methionine	1.50	0.0300
l-phenylalanine	3.30	0.0660
l-threonine	4.76	0.0952
l-tryptophane	0.80	0.0160
l-valine	4.68	0.0936
glycine	1.50	0.0300
l-serine	2.10	0.0420

Dissolve in one liter of triple distilled H_2O with a little warming. Freeze in 100 ml lots each for 5 liters of medium.

Vitamin Stock Solution for 50 Liters

Choline chloride	0.20	0.004
Nicotinamide	0.20	0.004
Calcium-d-pantothenate	0.20	0.004
Pyridoxal HCl	0.20	0.004
Thiamine HCl	0.20	0.004
Riboflavin	0.02	0.0004

Dissolve in 350 ml triple distilled H_2O; don't heat above 60°C.

Folic acid	0.20	0.004

Dissolve in 125 ml triple distilled H_2O by adding 1 N NaOH by drops until solution is clear. Add to the rest of the vitamins. Make up to 500 ml with triple distilled H_2O and distribute in 50 ml quantities (for 5 liters of medium). Store in deep freeze.

Ferric Nitrate Stock Solution

Fe(NO$_3$)$_3$·9 H$_2$O	0.010 gm
Triple distilled H$_2$O	100 ml

Distribute 50 ml in 5 ml quantities (for 5 liters of medium). Store in deep freeze. If necessary, use concentrated HCl to dissolve Fe(NO$_3$)$_3$·9 H$_2$O.

173

Table 12-2. Preparation of Dulbecco and Vogt's Medium

1. Thaw amino acid stock solution for 5.0 liters (100 ml).
2. Thaw vitamin stock solution for 5.0 liters (50 ml). If it is not completely dissolved, add 1 N NaOH by drops to bring into solution.
3. Dissolve 0.24 gm 1-cystine in 150 ml triple distilled H_2O by adding 1 N NaOH by drops with stirring.
4. Dissolve 0.36 gm 1-tyrosine in 300 ml triple distilled H_2O by heating to boiling.
5. Weigh and add to 6.0-liter flask:

NaCl	32.0 gm
KCl	2.0 gm
$CaCl_2$	1.0 gm (or $CaCl_2 \cdot 2H_2O$, 1.335 gm)

   Add 2 liters of triple distilled H_2O and dissolve.
6. Add the warm tyrosine solution.
7. Weigh and add to flask:

$MgSO_4 \cdot 7H_2O$	1.0 gm
$NaH_2PO_4 \cdot H_2O$	0.62 gm
Inositol	0.017 gm
Glucose	22.5 gm

8. Add $Fe(NO_3)_3$ solution (100 mg/liter), 5 ml.
9. Gas with 100% CO_2 for 3 or 4 minutes.
10. Add 18.5 gm of $NaHCO_3$ dissolved in one liter of triple distilled H_2O.
11. Add amino acids and vitamins to 7.5 ml of 0.5% phenol red to the cystine solution. Flush with 100% CO_2 to make the solution nearly neutral and add to the flask.
12. Add 2.9 gm of 1-glutamine.
13. Make up to 5.0 liters with triple distilled H_2O.
14. Gas to yellow color with 100% CO_2 before filtration and use 100% CO_2 for filtration pressure at 5 psi.
15. Filter and dispense. Incubate all bottles at 37°C for 48 hr to check sterility. After sterility check, antibiotics may be added.
16. Color code red for use on odd days and blue for use on even days of the month. Ideally, red and blue lots should be filtered at two separate times.

this acidity is lost during dispensing. It is useful to make aliquots of 170 ml so that 30 ml of serum can be added just before use (final serum concentration 15%). All media bottles should be incubated at 37°C for 48 hr and checked for turbidity prior to addition of antibiotics. (See section, this chapter, on antibiotics.) D & V is stored at refrigerator temperature, caps tightened, phenol red color on the yellow side. Serum should be stored frozen in aliquots to avoid repeated thawing. D & V medium requires equilibration with 10% CO_2 during culturing of cells. Glutamine may be added just prior to use.

BALANCED SALT SOLUTIONS (BSS)

Balanced salt solutions are needed to rinse cells prior to trypsinization, during processing for examination of chromosomes, and during other types of specialized procedures. pH in the range of 7.4, as well as isotonicity, should be maintained by these solutions. The protocol for making a useful balanced salt solution (PBS) is included here (Table 12-3). This solution is deficient in calcium and magnesium to aid in cell dispersion at the time of trypsinization.

Table 12-3. Composition of PBS (Phosphate Buffered Saline, Deficient in Ca and Mg)

16.0 gm NaCl
0.4 gm KCl
4.23 gm $Na_2HPO_4 \cdot 7H_2O$
0.4 gm KH_2PO_4

1. The reagents are diluted to 2000 ml with triple distilled H_2O and dissolved.
2. The solution is dispensed into 8-oz prescription bottles, 125 ml per bottle, and autoclaved.
3. Prepare two separate lots of PBS. Color code one lot red for use on odd days, and one blue for use on even days of the month.
4. The pH should be 7.40 $\pm$ .04. Check with pH meter.

Balanced Salt Solutions for Mammalian Cells
(available commercially)

This list of balanced salt solutions is not complete.

Earle's
Hank's
Phosphate buffered saline

SOLUTIONS TO DISPERSE MONOLAYER CELLS FOR SUBCULTURE
(Tables 12-4 and 12-5)

Trypsin is commonly used to disperse cells enzymatically from a monolayer into suspension. The strength and time of application of this enzyme are critical to insure good cell growth following trypsinization. (Also, see section, this chapter, on routine subculturing of monolayer cells.) Trypsin is inactivated by serum. Therefore the monolayer cells to be subcultured must be rinsed two times with balanced salt

Table 12-4.   Trypsin (0.25%)

1. Reconstitute freeze-dried trypsin (1:300, Hyland, in phosphate buffer with 1 mg Phenol Red) with 100 ml sterile triple distilled H_2O. After reconstitution, store frozen in sterile 10-ml plastic tubes, each containing 1 ml (2.5%). Color code red and blue before freezing.
2. For working solution (0.25%), add 9 ml sterile PBS (balanced salt solution).

solution before trypsinization.   After trypsinization the addition of medium containing serum prevents further action of the enzyme. Some types of human diploid cells, especially those of fetal origin, are more difficult to disperse. These cells will require trypsinization times up to 30 min, and trypsin concentrations up to 0.25%. The routine sub-culturing procedure outlined in this chapter employs the maximum trypsin strength (0.25%) because the tests for cell growth, such as single cell plating efficiency, mean cell cycle time, and cell viability by vital dye tests, reveal no damage to the cells from the maximum strength.

A trypsin-Versene (T-V) solution disperses the cells quickly and completely. Versene (EDTA) binds calcium and magnesium and therefore increases cell dispersion. However, dispersion by T-V solution is recommended only for "dead-end" experiments and cell counts. If T-V solution is used for routine subculturing of human cells, the culture will gradually fail, since Versene is not inactivated or removed completely.

Dispersal of the cells from monolayer attachment by mechanical means, such as scraping or shaking, may be indicated in certain situations.

Table 12-5.   0.25% Trypsin-0.25% Versene (T-V Solution) (For Cell Counts, Not for Routine Subculturing)

1. Store 1 ml frozen, sterile aliquots of 2.5% trypsin.
2. Store 1 ml frozen, sterile aliquots of 2.5% Versene (EDTA, ethylene-diaminetetraacetic acid, disodium salt).
3. For working solution (0.25% trypsin-0.25% Versene), add 1. and 2. above to 8 ml sterile PBS (balanced salt solution).

ANTIMICROBIAL AGENTS

The addition of antimicrobial agents to tissue culture medium is optional. Many investigators prefer to leave them out, since they may slow down detection of a partially sensitive contaminant. If used, they should be added to the medium after it has been filtered and sterility has been checked by incubation at 37°C for 48 hr. The antibiotics, penicillin and streptomycin, are commonly added together (Tables 12-6 and 12-7).

Antifungal agents may be added to medium for limited periods of time. Nystatin (Mycostatin, Squibb) is described here (Table 12-8).

Table 12-6. Penicillin (Sodium) (100 units/ml, Final Concentration)

1. Sterile vial containing 1,000,000 units is diluted with 10 ml sterile triple distilled H$_2$O.
2. To dispense per 200 ml of medium in individual bottles, add 0.2 ml to each bottle. (100 units/ml, final concentration of penicillin.)
3. Mark Ab or some other symbol on label.

Table 12-7. Streptomycin (0.1 mg/ml, Final Concentration)

1. Vial containing 1 gm Streptomycin Sulfate (Squibb) is diluted with 5 ml sterile triple distilled H$_2$O.
2. To dispense per 200 ml of medium in individual bottles, add 0.1 ml to each bottle. (0.1 mg/ml, final concentration of streptomycin.)
3. Mark Ab or some other symbol on label.

Table 12-8. Nystatin (25 units/ml, Final Concentration)

1. Sterile vial containing 500,000 units is diluted with 10 ml sterile triple distilled H$_2$O.
2. Add to a final concentration of 25 units/ml (0.25 ml/500 ml medium).

ROUTINE HANDLING OF STOCK MONOLAYER CULTURES

Routine subculturing is also called "farming." A record book should be kept which describes each subculture according to how the bottle is divided. The exact procedures are described in Chapter 5 under long-term culture of human cells. If the division is in half, the process is referred to as a 1:2 split, which represents 1 population doubling. A 1:4 split is 2 population doublings; a 1:8 split is 3 population doublings, and so on. It is now generally agreed that the age of a culture should be described by the number of population doublings rather than by the number of subcultures or time in culture. A suggested format for record keeping is presented in Table 12-9.

Every culture should be entered in the first part of the record book and assigned a number, in order as the specimen reaches the laboratory. Other pertinent data are also recorded in this section. (See Chapter 13 on tissue culture nomenclature.) Pertinent data include source, name, age, hospital number (if a patient), date received.

The routine subculture record (Table 12-9) does not take into account the number of doublings required during the time between the

Table 12-9. Format for Record of Serial Tissue Culture

No. of Doublings	Bottle No.	Source, Date	Split	Medium Change, Date	Subculture Into:	Discard Date (Note why, if other than subculture)
0	25-1	Primary explant 5-1-68		5-3-68	25-2,3	5-6-68
1	25-2	25-1 5-6-68	1:2		25-4,5,6,7	5-11-68
1	25-3	25-1 5-6-68	1:2	5-12-68		5-25-68 Cells old
3	25-4	25-2 5-11-68	1:4			5-16-68 Cells frozen
3	25-5	25-2 5-11-68	1:4			5-16-68 Cells frozen
3	25-6	25-2 5-11-68	1:4		25-8,9	5-16-68
3	25-7	25-2 5-11-68	1:4	5-17-68		
5	25-8	25-6 5-16-68	1:4			5-20-68 Expt. #3
5	25-9	25-6 5-16-68	1:4			
			and so on			

178

first primary explant outgrowth and the first subculture. This number of doublings is determined by the number of original cells from the explant and the number of cells in the container used for the first subculture. A general estimate is in the range of 20 doublings. A more exact estimate can be made in each individual case, if necessary. Because of the limited *in vitro* lifetime of human diploid cells (50 ± 10 doublings) it may be important to know the culture age fairly exactly in order to predict how long the culture can be used. If cells are frozen for storage in liquid nitrogen as early as possible in the lifetime of the culture, older cells in culture can be discarded and younger cells retrieved from storage. (See section, this chapter, on long-term storage of tissue culture cells.)

Reference: Hayflick, L.: The Limited *in Vitro* Lifetime of Human Diploid Cell Strains. Exp. Cell Res., *37*:614, 1965.

SUSPENSION CULTURES

As yet, human diploid cells are not commonly grown in serial suspension culture. The HeLa established cell line was originally started from a human cervical carcinoma, is far from diploid, and can be grown as either a monolayer or a suspension culture. Other nondiploid mammalian established cell lines, such as Chinese hamster ovary and mouse L-cell, can be grown in suspension and carried serially. Cells in suspension can be cultivated in large volumes and can be sampled at intervals without disturbing the remainder of the cells in the container. Suspension cultures must be shaken, agitated, or rolled to keep the cells from coming out of suspension. (See one of the tissue culture texts recommended on page 166 for further details of maintenance.)

LONG-TERM STORAGE OF TISSUE CULTURE CELLS

Tissue culture cells can be suspended as single cells, frozen at a controlled rate of 1°C per min, and stored for prolonged periods of time at the temperature of liquid nitrogen (−320°F at atmospheric pressure).

Equipment

Liquid nitrogen refrigerator (LR-35; 6 or 9 canister sizes; Linde Co., N. Y.; satisfactory for a moderate amount of cell storage for one tissue culture laboratory; larger, smaller, and newer models are available).

179

Biological freezer (BF-3; includes freezing chamber and controller unit; Linde Co., N. Y.).

Electronic recorder (Electronik 18, Honeywell, Philadelphia, Pa.).

Liquid nitrogen storage tank (available from local supplier of compressed gas).

Preparation of Cells for Freezing

1. Trypsinize monolayer cells in the usual manner for subculturing. (See previous section, this chapter, on routine handling of stock cultures, and also Chapter 5.) Cells should not be overly confluent.

2. When cells are removed from the bottom surface, add medium and suspend well. They should be single, and without clumps.

3. Centrifuge at 600 to 800 rpm for 10 min.

4. Remove supernatant.

5. Resuspend in medium with 10% final concentration of sterile glycerin.* Amount of medium should be adjusted to make the cell count not less than 2×10^6/ml.

6. Add 2 ml of cell suspension to labeled sterile freezing ampules (Wheaton Gold Band "Cryules LN," vitro "200," prescored, for preservation of biological materials with liquid nitrogen, 2-ml size, Wheaton Laboratory Ware, Millville, N. J.).

7. The technique of sealing the ampules is very important. If a small leak is present, liquid nitrogen can enter the ampule when it is frozen, and the ampule will explode when it is thawed, due to rapid expansion of the liquid nitrogen. Some investigators place the sealed ampules in a dye solution prior to freezing, to check for leaks.

   a. Use a blowpipe and burner combination with medium to large round tip. (Special ampule sealing burners are available but are less useful for other general laboratory procedures.)

   b. Hold ampule in one hand and glass stirring rod in the other. Be sure the neck of the ampule is free of liquid. Touch the rod to the tip of the ampule and put the two together into the blue tip of the flame.

   c. When both rod and ampule are red hot and the rod sticks to the ampule, pull apart gently and twist the end of the ampule to seal it.

---

* Glycerin may be autoclaved.

    d. Allow ampule to cool.

    e. Scrape off excess glass bits from the stirring rod on a wire screen before using again.

Freezing Cells

Detailed instructions accompany the freezing equipment, and these instructions should be read carefully. An electronic recorder records the rate of cooling of the cells in the freezing chamber. A separate controller unit regulates the entry of liquid nitrogen from the main tank into the freezing chamber. The controlled drop of $1°C/min$ should be followed to $-40°C$; then the temperature may be reduced quickly to $-80°C$. At this point the ampules are transferred rapidly to racks which are inserted into the canisters in the liquid nitrogen refrigerator. (See the next section on storing cells.)

Storing Cells*

1. If the culture number is written on the freezing ampule with a standard glass marking pencil (not wax), the label remains readable after retrieval from liquid nitrogen. The end of the rack holding the ampules should also be labeled with the culture number. (Aluminum racks which each hold three 2-ml ampules can be obtained from Frozen Seamens, Division of Cryo-Therm Inc., Breinigsville, Pa.) It is unwise to place more than one type of culture on one rack. Furthermore, if ampules are not on racks, they are difficult to retrieve from the storage refrigerator.

2. An accurate and up-to-date record of what cells are stored and where they are stored is important. Table 12-10 is a suggested record for cell storage. If cells of one culture number are frozen at different population doublings, a subletter is assigned to each new freeze (Example: 105-d). When an ampule is removed, the date and name of the person using it should be noted after the ampule number. If the ampule is cultured and the cells are refrozen, make a note of the new freeze subletter after the ampule number (Example: 105-d→ 105-e on 6-18-68).

3. The level of liquid nitrogen should be checked and more added at appropriate intervals. It is important to store the cells in the liquid phase.

---

* The storage described here is for the Linde LR-35 refrigerator. Other types of refrigerators are also available. Which type to use depends on the needs of the laboratory and the volume of cells to be stored.

Table 12-10. Suggested Record for Cell Storage

Culture Number:						
Description of Culture:						
(See Chapter 13 on tissue culture nomenclature)						
Freeze Letter. Ampule Number	*Refrigerator Compartment Number*	*Number of Cells/ml*	*Date of Freezing*	*Number of Ampules Frozen*	*Medium*	*Age of Cells (Doublings)*
a	2A2B	2×10^6	4-1-68	3	D&V 15% FC	25
1						
2						
3						
b	3A3B	2×10^6	5-10-68	3	D&V 15% FC	30
1						
2						
3						

4. It is wise to keep the refrigerator in a cold room to retard cold loss.
5. Heavy insulated gloves should be worn at all times when racks or canisters are handled. The room should be well ventilated when the refrigerator is opened, and extreme care should be taken to avoid spilling liquid nitrogen on the skin.

Thawing Cells

1. Know exactly which ampule is needed and where it is.
2. A stainless steel pan and metal thermometer are used to obtain water at 37°C. Place this pan near the refrigerator. It is also wise to place a shield over the pan, as well as over the face of the person removing the cells, in case an ampule explodes. Heavy plastic face and table shields can be purchased or made.
3. Remove the ampule as quickly as possible and place in water bath.
4. Thawing occurs very rapidly, within 1 min.
5. Place 1 or 2 ml of cell suspension in 10 ml of medium for culture.
6. The medium may be changed after 24 to 48 hr if a number of floating cells are present.

TISSUE CULTURE CONTAINERS—DISPOSABLE

Prescription bottles—1 oz, 2 oz, 4 oz, 8 oz, 16 oz, 32 oz.

Rinse with triple distilled H_2O, place caps on loosely, cover with aluminum foil and autoclave. (See section on sterilization, this chapter.) Teflon liners may be placed inside the caps to make them gastight (Arthur H. Thomas Co., Philadelphia, Pa.). Use #18 or #20 for 1-oz, #22 for 2- and 4-oz, #24 for 8-oz, and #28 for 16-oz bottles.

Tissue culture plastic containers, sterile (Falcon Plastics, Los Angeles, Calif.).

Petri dishes—100 $\times$ 20, 60 $\times$ 15 mm.

Flasks—250 ml, 30 ml.

SUPPLIES AND EQUIPMENT NECESSARY FOR TISSUE CULTURE

Sterile Disposable Supplies

Plastic tubes, sterile, with snap caps, various sizes, individually wrapped.

Plastic pipettes, sterile, with cotton plugs, serological, 1 ml, 5 ml, 10 ml.

Pasteur pipettes, long tips, sterilized by dry air (see page 166).

Syringes, 1 ml, 10 ml, 30 ml, sterile.

Needles, #18, #20, #22, sterile.

Sterile Permanent Supplies

(Wash and sterilize by instructions, this chapter.)

5-ml, 10-ml serological pipettes, constricted mouth to hold cotton plugs, sterilize in pipette cans.

Rubber bulbs, 1-ml capacity; rubber policemen for scraping off cells, etc.

See tissue culture glassware catalogue for test tubes, Petri dishes, storage cans, and tissue culture containers as needed.

Desiccators, various sizes.

Instruments—scissors, forceps, etc.

Supplies—Other

Funnels, cylinders, flasks, plastic storage bottles.

Stainless steel pans and baskets for sterilization.

Plastic jars and baskets to hold pipettes for soaking and washing.

Miracloth (lint-free cloth, Chicopee Manufacturing Corp., Milltown, N.J.).

Aluminum foil.

Wescodyne (detergent-germicide, West Chemical Products Inc., Long Island City, N. Y.).

7X (detergent, Linbro Chemical Co., New Haven, Conn.).

Clorox bleach.

Test tube brushes, various sizes; scouring pads.

Test tube racks.

Silicone grease.

Sterilizing tape (dry air and autoclave).

Centrifuge tubes, heavy duty, some should be graduated.

Small Equipment

Hemocytometer

Filters and filter holders of various sizes, to fit on syringes, or to filter by suction or positive pressure. (See Millipore or Gellman catalogues. Large filter holders with containers to receive and deliver fluid volumes of more than one liter are major equipment items.)

Lab timers

Lab cart

Bunsen burners

100% CO_2 tank

5% or 10% CO_2 in compressed air tank $\Big\}$ with pressure gauges

Major Equipment

Water-jacketed incubator

Water bath

Culture hood

Inverted light microscope

Standard light microscope

Pipette washer, pipette rinser

Drying and sterilizing oven

Autoclave

Water still

Centrifuges, table and floor models

Analytic balance

Standard refrigerator and freezer

See section, this chapter, for description of equipment for freezing and long-term storage of cells.

13

Tissue Culture Nomenclature

Terminology in the area of tissue culture has been confusing because different laboratories have used words in different ways. The final draft of a proposed usage of animal tissue culture terms was accepted by the Tissue Culture Association in 1966 and is the basis for the definitions in this chapter.

Reference: Committee on Terminology, Tissue Culture Association: Proposed Usage of Animal Tissue Culture Terms. Cytogenetics, 6:161, 1967.

TISSUE CULTURE, DEFINITION

Animal tissue culture is concerned with the study of cells, tissues, and organs explanted from animals and maintained or grown *in vitro* for more than 24 hours. Dependent upon whether cells, tissues, or organs are to be maintained or grown, two methodological approaches have been developed in the field of tissue culture.

1. *Cell culture.* This term is used to denote the growing of cells *in vitro,* including the culture of single cells. In cell cultures the cells are no longer organized into tissues.

186

2. *Tissue or organ culture.* This term denotes the maintenance or growth of tissues, organ primordia, or the whole or parts of an organ *in vitro* in a way that may allow differentiation and preservation of the architecture or function.

TYPES OF CULTURES

Explant. This term describes an excised fragment of a tissue or an organ used to initiate an *in vitro* culture.

Monolayer. The term refers to a single layer of cells growing on a surface.

Suspension culture. This term denotes a type of culture in which cells multiply while suspended in medium.

Primary culture. This term implies a culture started from cells, tissues, or organs taken directly from organisms. A primary culture may be regarded as such until it is subcultured for the first time. It then becomes a "cell line."

Cell line. A cell line arises from a primary culture at the time of the first subculture. The term *cell line* implies that cultures from it con-

187

sist of numerous lineages of the cells originally present in the primary culture.

Established cell line. A cell line may be said to have become "established" when it demonstrates the potential to be subcultured indefinitely *in vitro.*

Cell strain. A cell strain can be derived either from a primary culture or a cell line by the selection or cloning of cells having specific properties or markers. The properties or markers must persist during subsequent cultivation. In describing a cell strain its specific feature should be defined—*e.g.,* a cell strain with a certain marker chromosome, a cell strain that has acquired resistance to a certain virus, or a cell strain having a specific antigen.

Clone. This term denotes a population of cells derived from a single cell by mitoses. A clone is not necessarily homogeneous, and therefore the terms *clone* or *cloned* should not be used to indicate homogeneity in a cell population.

Cloned strain or line. This term denotes a strain or line descended directly from a clone. (See *clone.*)

Diploid cell line. This term denotes a cell line in which, arbitrarily, at least 75% of the cells have the same karyotype as the normal cells of the species from which the cells were originally obtained. A description of a diploid cell line should include the actual numbers of cells examined, the percentage of diploid cells, and their karyotype.

Heteroploid cell line. This term denotes a cell line having less than 75% of cells with diploid chromosome constitution. This term does not imply that the cells are malignant or that they are able to grow indefinitely *in vitro.* In describing a heteroploid cell line, in addition to the karyotype of the stem line, the percentage of cells with such karyotype should be stated.

CHROMOSOME NUMBERS

Haploid. 1. The basic number of a polyploid series (symbol: x). Haploid in this meaning = monoploid.

2. The chromosome number of the haplophase, the gametic, reduced number (symbol: n).

Diploid, triploid, tetraploid, etc. The double, triple, quadruple, etc., basic number (symbols: 2x, 3x, 4x, etc.).

188

Polyploid. General designation for multiples of the basic number, higher than diploid.

Heteroploid. 1. In organisms with predominating diplophase: all chromosome numbers deviating from the normal chromosome number of the diplophase.

2. In organisms with predominating haplophase: all chromosome numbers deviating from the normal chromosome number of the haplophase.

Euploid. All exact multiples of x.

Aneuploid. All numbers deviating from x and from exact multiples of x.

Mixoploidy. The presence of more than one chromosome number in a cellular population.

Endopolyploidy. The occurrence in a cellular population of polyploid cells, which have originated by endomitosis.

OTHER TISSUE CULTURE TERMS

Subculture. This term denotes the transplantation of cells from one culture vessel to another.

Subculture number. This term indicates the number of times cells have been subcultured, *i.e.,* transplanted from one culture vessel to another.

Subculture interval. This term denotes the interval between subsequent subcultures of cells. This term has no relationship to the term *cell generation time.*

Cell generation time. This term denotes the interval between consecutive divisions of a cell. This interval can be best determined at present with the aid of cinematography. This term is not synonymous with *population doubling time.*

Population doubling time. This term is used when referring to an entire population of cells and indicates the interval in which, for example, 1×10^6 cells increase to 2×10^6 cells. This term is not synonymous with *cell generation time.*

Absolute plating efficiency. This term indicates the percentage of individual cells that give rise to colonies when inoculated into culture vessels. The total number of cells in the inoculum, the type of culture vessel, and the environmental conditions (medium, temperature, closed or open system, CO_2 atmosphere, etc.) should always be stated.

Relative plating efficiency. This term indicates the percentage of inoculated cells that give rise to colonies, relative to a control in which the absolute plating efficiency is arbitrarily set as 100. The total number of cells in the inoculum, the environmental conditions, and the absolute plating efficiency of the control should always be stated.

Fibroblasts. Fibroblasts are cells of spindle or irregular shape and, as their name implies, are responsible for fiber formation. In cell cultures many other cell types are morphologically indistinguishable from fibroblasts. In organ and tissue cultures, in which cell interrelationships are preserved, fibroblasts may be identified by accepted histological criteria.

Fibroblast-like cells. In cell cultures various types of cells acquire similar morphology. Cells acquiring irregular or spindle shape are often referred to as "fibroblasts." However, the derivation of these cells or their potentialities, such as production of fibers, are usually not known. Therefore, such cells are more properly called "fibroblast-like" cells.

Epithelial cells. This term refers to cells apposed to each other, forming continuous mosaic-like sheets with very little intercellular substance, as seen in *in vivo* or in tissue or organ cultures.

Epithelial-like cells. In cell cultures epithelial cells may assume various shapes but tend to form sheets of closely adherent polygonal cells. However, the degree of cohesion of the cells can vary. When the only criterion for identification of such cells is their tendency to adhere to each other, it is preferable to refer to the cells as "epithelial-like" cells.

Culture alteration. This term is used to indicate a persistent change in the properties or behavior of a culture—*e.g.,* altered morphology, chromosome constitution, virus susceptibility, nutritional requirements, proliferative capacity, malignant characters, etc. The term should always be qualified by a precise description of the change that has occurred in the culture. The term *cell transformation* should be reserved to mean changes induced in the cells by the introduction of new genetic material. The nature and source of the genetic material inducing the change should be specified.

INFORMATION USED TO DESCRIBE A NEW CELL LINE

1. Whether the tissue or origin was normal or neoplastic and, if neoplastic, whether benign or malignant.
2. Whether the tissue was adult or embryonic.

3. The animal species of origin.
4. The organ of origin.
5. The cell type (if known).
6. The designation of the line.
7. Whether the line has been cloned.

DESIGNATION OF A CELL LINE

1. Not more than four letters in series indicating the laboratory of origin.
2. A series of numbers indicating the line.
   Example: NCL 123

INFORMATION USED TO DESCRIBE A CELL LINE OR A CELL STRAIN FOR PUBLICATION

1. History.
2. Population doubling number (Subculture number).
3. Culture medium.
4. Growth characteristics.
5. Absolute plating efficiency.
6. Morphology.
7. Frequency of cells with various chromosome numbers in a culture.
8. Karyotype(s) characteristic of the stem line(s).
9. Whether sterility tests for Mycoplasmas, bacteria, and fungi have been done.
10. Whether the species of origin of the culture has been confirmed and the procedures by which this was done.
11. Virus susceptibility of a culture at a given subculture number.
12. A description of a cell strain should also include the procedure of isolation, the specific properties of the cells in detail, the number of population doublings, and the length of time since isolation.

14

The Handling of Primary Explants For Long-Term Cell Culture

___

This chapter is devoted to the establishment of *long-term cell cultures*. Organ cultures are not included. It is important to refer to Chapter 12 for other principles of tissue culture. Procedures that include cell culture from a *primary explant* but conclude before the first subculture are discussed in Chapter 6 on the analysis of chromosomes directly from tissues.

> *References:* Paul, J.: *Cell and Tissue Culture.* 3rd ed. Baltimore, The Williams and Wilkins Co., 1965, Ch. XI.
> White, P. R.: *The Cultivation of Animal and Plant Cells.* 2nd ed. New York, The Ronald Press Co., 1963, Ch. 6.

HUMAN SKIN BIOPSIES

Skin is perhaps the easiest human tissue to sample for the establishment of long-term cultures unless tissues are available from surgical procedures for other reasons. Every laboratory has slightly different details in the skin biopsy protocol. General principles are the same, and involve (1) sampling of dermis so that fibroblast-like cells may grow out in long-term culture; (2) a method to anchor the primary explant in a culture container; (3) proper growth conditions; and (4)

192

patience, since under optimal conditions human diploid fibroblast-like cells require 3 to 6 weeks before the first subculture can be attempted, and then the cells may grow slowly.

Reference: Harnden, D. G., and Brunton, S.: The Skin Culture Technique. In: *Human Chromosome Methodology,* J. J. Yunis, ed. New York, Academic Press, 1965.

The Pinch Skin Biopsy Technique

1. Scrub the skin well with 70% ethanol or with 0.02% iodine in 70% ethanol (1 ml 2% iodine tincture in 99 ml 70% ethanol), followed by 70% ethanol. The inside of the forearm is a good area for persons of all ages because this area is less sensitive to pain, is easily accessible, and is free of hair.

2. With a long-tipped sterile forceps obtain a pinch of skin about 10 mm long and 1 to 1.5 mm wide. Apply increasing pressure with the forceps and hold until the skin becomes white. The individuals will feel uncomfortable pressure. It is best to ask them not to watch and proceed quickly with the skin excision at

193

this point in the procedure. The pinch is reasonably good anesthesia for the excision.

3. Use a sterile scissors with fine blades. Snip quickly along the 2 points of the forceps. The specimen should be about 4 to 5 mm long and 1 mm wide. The exact size will vary from one person to another. The ease of obtaining dermis will also vary but will be uniformly accomplished if the pinch and excision are as large as just described. Pinpoint bleeding at the site of the excision is an indication that mesodermal tissue has been obtained.

4. Drop the specimen immediately into a small Petri dish containing balanced salt solution. Process within several hours (up to 24 hr).

5. Lift up the tissue with a fine forceps and cut off pieces smaller than 1×1 mm. Sterile disposable needles may be used to transfer the small fragments to the culture container.

6. Sterile 30-ml plastic tissue culture flasks (Falcon Plastics) are excellent for starting primary explants for serial culture. The fragments can be inserted through the narrow neck and placed as far in as desired. Include 1 to 3 tissue pieces in each container. At least 2 and preferably 3 containers are recommended.

7. Allow the fragments to dry out *briefly*. During this period the tissue sticks slightly to the plastic surface.

8. Carefully let 1 drop of culture medium plus 20% to 25% fetal calf serum fall from a Pasteur pipette over each fragment. On the plastic surface, a drop will "bead" and will serve the function of a "clot" to hold the fragment. About 0.25 to 0.5 ml of medium may also be added directly to another portion of the flask for humidification, but care must be taken to keep this amount from running into the area of the fragments.

9. Gas gently to the proper pH and cap tightly. Also place in a dry desiccator or gas flow incubator at the proper concentration of CO_2. This additional precaution insures proper pH even if the cap of the culture container leaks. Primary explants are extremely sensitive to changes of pH.

10. Incubate at 37°C for about 2 days, undisturbed. Look at the color of the medium, if possible without moving the culture containers. Only if the pH is wrong should the cultures be manipulated at this time. Re-gas if necessary.

11. After this period, the flasks may be checked under an inverted microscope for signs of evaporation or undue amounts of pre-

cipitated material in the medium. Several more drops of medium may be added with a Pasteur pipette.

12. When new epithelial-like or fibroblast-like growth first appears at the edge of the primary explant (after about 4 days, with a range of about 2 to 10), the explant is less susceptible to "floating" away from the surface. Nevertheless, care must still be taken when medium is added. Thirty-ml plastic tissue culture flasks (Falcon Plastics) will hold about 3 ml. Subsequently change medium every 4 to 5 days.

13. Do not be in a hurry to make the first subculture. A good halo of fibroblast-like cells covering an area larger than the primary explant should be present. Some explants will show rings of epithelial-like growth before the fibroblast-like cells appear. These epithelial-like cells cannot be subcultured by the usual procedures for monolayer fibroblast-like cells.

14. Cells from the first trypsinization may be transferred to a 250-ml plastic flask or 8-oz prescription bottle in about 10 to 15 ml of medium plus 15% or 20% serum. Gently run 1 ml of balanced salt solution (PBS, see Chapter 12) over the primary explant, using a Pasteur pipette. Aspirate. Discard fluid. Repeat 2X. Add 0.25 to 0.5 ml of 0.25% trypsin, making sure the area of fibroblast-like growth is covered. Incubate at 37°C until the cells detach (not longer than about 20 min).

15. Rinse 1 ml of medium gently over the cells. Aspirate and add to the new bottle. Repeat several times. Add new medium to the primary explant.

16. Return cultures to the incubator at proper pH and humidified 37°C.

17. Allow the first subculture to reach near-confluency and subculture to two bottles. Keep a culture record as described in Chapter 12.

18. Consider freezing a portion of the cells as soon as possible for prolonged storage in liquid nitrogen.

19. Primary explants may be retrypsinized at intervals after new outgrowth occurs.

Method of: Priest J. H.: Unpublished.

Modifications

1. The primary explant may be anchored under a coverslip in a 60-mm sterile plastic tissue culture Petri dish (Falcon Plastics),

or with care the fragments may be grown in the Petri dish without any special attachment method.

2. The Leighton tube method may be used. (See page 78.)
3. The primary explant may be placed in a clot that holds it and also furnishes some nutrition. Our experience with chicken plasma and chick embryo extract (obtained commercially) is that when mixed, they usually do not clot. Otherwise, human primary explants do reasonably well in these clots plus tissue culture medium supplemented with 20% fetal calf serum.
4. Various types of skin biopsy devices have been designed to obtain a specimen of uniform size and content at fast speed. In our opinion, the pinch biopsy described in detail here is equally reliable and reproducible.
5. The epithelial-like outgrowth, although it cannot be maintained in serial culture, may show new rings of growth each time the primary explant is trypsinized. Eventually the increase in size stops, and the central rings deteriorate and lift from the surface. This epithelial-like growth can be shown to have differentiated epithelial function not maintained by the fibroblast-like cells (*i.e.,* histidase activity).
6. Chromosome examination may be performed on the initial cellular outgrowth before the first subculture. (See page 78.)
7. *In vitro* trypsinization of the primary explant may be used prior to culture. (See Chapter 6.)

The Cellophane Strip Technique

1. Take roll of perforated cellophane (Microbiological Associates, Bethesda, Md., and Albany, Calif.) and cut it into strips of proper size to fit into the bottom of the culture flask. The cellophane will have a relatively smooth and a relatively rough side.
2. Cut off the same corner from each strip to mark them.
3. Wash strips successively overnight in ether, acetone, and 95% alcohol.
4. Boil strips in several changes of distilled water.
5. Insert strip into clean culture flask with the rough side of the cellophane against the glass.
6. Pipette in enough water to cover the cellophane strip in order to soften it.
7. Autoclave flask with cellophane in it.

8. When ready to use flask, pour off water and rinse flask with sterile media.
9. Rotate flask 90° and insert tissue explants onto bare floor of flask.
10. With sterile curved forceps pull cellophane onto tissue explants. (Smooth side will now be down on top of explants.)
11. Gently feed with medium.

Steps 3 and 4 are elective.

Method of: Hecht, F., and Jentoft, V.: University of Oregon Medical School, Portland, Oregon.

HUMAN SURGICAL SPECIMENS, AUTOPSY TISSUES, FETAL TISSUES

Surgical Specimens

1. Specimens must be handled with sterile technique and processed as quickly as possible after removal from the patient.
2. In general, they may be established according to the method, or its modifications, described in detail for skin biopsies. Several rinses in balanced salt solution may be indicated particularly to remove red blood cells.
3. In general, fibroblast-like cells will grow out, and chromosome complement of the primary explant will be maintained throughout the lifetime of the human cell culture (50 ± 10 population doublings). An occasional exception is the culture from a human tumor which may have an unstable chromosome complement in culture.
4. Dense tissues may require *in vitro* trypsinization as described for the examination of chromosomes from solid tumors. (See Chapter 6.)
5. Foreskin is an excellent source from which to establish long-term cultures. The skin preparation recommended for skin biopsies is also recommended for circumcisions to be used for tissue culture.

Autopsy Tissues

1. The chance of contamination increases. Autopsy skin may be prepared with iodine and alcohol as described for skin biopsies of living tissues. Internal organs should be dissected with sterile instruments, or else the interior of large organs may be dissected

197

out by sterile technique. Only small pieces of tissue are needed for culture. Follow the rules already described for skin biopsies. Include several rinses in balanced salt solution.

2. Thymus in children makes a good primary explant. Small round cells will be present initially before the fibroblast-like cells overgrow the culture. Initiation of the monolayer fibroblast-like culture may be shorter for thymus than for other tissues.

3. Gonad cultures may be indicated in cases of sex chromosome malformations, intersexes, or chromosomal mosaicism. Again, the fibroblast-like cell will grow. Mitotic chromosomes are obtained, not meiotic chromosomes.

4. At autopsy, multiple tissue sampling may be performed for the evaluation of chromosomal mosaicism.

Fetal Tissues

1. The same culturing principles apply as for skin biopsies. In general, the new outgrowth occurs more rapidly.

2. Once fetal tissues are established, they may grow more rapidly (slightly shorter mean cell cycle time), although this rule is not absolute. The culture life is longer by about 10 population doublings.

3. Fetal fibroblast-like cells may be slightly more difficult to trypsinize and obtain as single cell suspensions for subculture than are cultures of adult tissue origin. Again, this rule is not absolute.

4. For studies of the chromosomes in abortions, refer to:

Geneva Conference: Standardization of Procedures for Chromosome Studies in Abortion. Cytogenetics, 5:361, 1966.

HUMAN AMNIOTIC FLUID

1. Centrifuge the amniotic fluid at 600 to 800 rpm for about 8 min and resuspend the pellet in culture medium with 15 to 30% added serum. (See Chapter 12.)

2. Place about 4 ml of cell suspension in each of 4 (or fewer) 60-mm plastic tissue culture Petri dishes.

3. Incubate at 37°C, humidified, at proper pH.

4. Check Petri dishes under an inverted microscope for cell attachment and multiplication of fibroblast-like monolayer.

Reference: Steele, M. W., and Breg, W. R., Jr.: Chromosome Analysis of Human Amniotic-Fluid Cells. Lancet, *i:*7434, 1966.

198

Modifications

1. Type of medium and serum are critical for the growth of cells at low density.
2. Some investigators suspend the cell button in a small amount of fetal calf serum in 60-mm Petri dish, cover with coverglass fitted to dish, incubate at 37°C for 1 hr, and add medium to desired concentration (not more than 30%).

CLONAL GROWTH FROM PRIMARY EXPLANTS

The techniques for clonal growth from primary explants are changing rapidly and differ considerably from one laboratory to another. There is no question that primary cloning, or the preparation of a culture derived from a single cell of a primary explant, has many uses in cell culture research laboratories. In some protocols the preparation of single cells is less strictly followed, and the "clones" are therefore not derived from single cells.

References: Ham, R. G., and Murray, L. W.: Clonal Growth of Cells Taken Directly from Adult Rabbits. J. Cell. Physiol., *70:*275, 1967.

Pious, D. A., Hamburger, R. N., and Mills, S. E.: Clonal Growth of Primary Human Cell Cultures. Exp. Cell Res., *33:*495, 1964.

Senn, J. S., McCulloch, E. A., and Till, J. E.: Comparison of Colony-Forming Ability of Normal and Leukemic Human Marrow in Cell Culture. Lancet, *ii:*597, 1967.

Wu, A. M., Siminovitch, L., Till, J. E., and McCulloch, E. A.: Evidence for a Relationship between Mouse Hemopoietic Stem Cells and Cells Forming Colonies in Culture. Proc. Nat. Acad. Sci., USA, *59:*1209, 1968.

15

Special Procedures On Cells in Long-Term Culture

QUANTITATIVE PLATING OF SINGLE CELLS

The quantitative plating of single cells is used to check the plating efficiency of medium or serum and to clone single cells (Tables 15-1 and 15-2). Cell counts are performed in a standard hemocytometer.

CLONING PROCEDURES

Ring Method from Single Cell Platings

The procedure described for single cell platings is a modification of the ring method originally described by Puck, Marcus and Cieciura.

1. When clones have formed from single cell platings (after 1 to 2 weeks, depending on the growth rate of the cells, see Table 15-2), select the number of clones to be subcultured. Examine the plate under an inverted microscope. Select clones that are isolated, compact, and composed of approximately 100 or more cells. With a wax pencil, place a ring on the bottom of the Petri dish to mark the exact position of the clone. More than one clone may be taken from each dish.

2. Decant the medium from the Petri dish, add 5 ml balanced salt solution (PBS), and pour off; repeat the rinse. Aspirate the last drops of PBS with a Pasteur pipette.

3. Using sterile technique and a round wooden applicator, place silicone grease on one side of a cloning ring. This ring should have an outside diameter of about 1.5 cm and an inside diameter of about 0.5 cm (the dimensions may be smaller but not larger). Height should be about 1.0 cm. The rings may be cut from solid teflon rods in a machine shop. The material used for the rings should withstand autoclaving.

4. Place ring over the wax mark on bottom of Petri dish, using a sterile forceps. Press the ring down firmly. If too much silicone grease is used, it will spread over the clone and prevent removal of the cells. Check placement of the ring under an inverted microscope. The clone should be centered in the ring.

5. Place 1 drop of 0.25% trypsin from a Pasteur pipette into the center of the ring. Incubate about 2 min at 37°C or until the cells appear rounded through the microscope. (Trypsin may also be left on about 10 min at room temperature.) Longer

Table 15-1. Plating Efficiency of Serum

Record control or lot No. of serum to be checked _____.

Date plates set up _____.

Inoculate 10 60-mm plastic tissue culture Petri dishes with 100 cells each. Use 15% serum in D & V or other culture medium for mammalian cells. (Other concentrations of serum may be substituted.)

Procedure (for 8-oz Prescription Bottles)

Select a bottle of cells from 75 to 100% confluent. Pour off medium. Add 2 to 3 ml balanced salt solution (PBS). Rotate. Pour off. Add 2 to 3 ml PBS. Rotate. Pour off. Add 1 ml 0.25% trypsin. Incubate until cells come off. Add 4 to 5 ml of medium plus 15% serum. Suspend cells well with Pasteur pipette and perform cell count (use both sides of counting chamber).

Counts per Large
Corner Squares   8 squares   Divide by 8 $\times 10^4 =$
1 side of chamber   cells/ml $= n$

1   2   3   4

Other side of chamber

1   2   3   4

$$\frac{n}{1} = \frac{1 \times 10^6}{ml}$$

Use 10-ml dispo-plastic tubes and 1-ml dispo-plastic pipettes to measure cell suspension.

Add _____ ml of cell suspension to make 10 ml in medium plus 15% serum (1×10^6 cells/10 ml).

Add 0.2 ml of cell suspension to 9.8 ml medium plus 15% serum (2×10^4/10 ml).

Add 1.0 ml of cell suspension to 9.0 ml medium plus 15% serum (2×10^3/10 ml).

Be sure cells are well suspended before each dilution.

To each Petri dish containing 4 ml of medium plus 15% serum add 0.5 ml of cell suspension (100 cells per dish). Immediately after addition of cells, agitate Petri dish to insure good distribution. Place dishes in incubator at 37°C for one week undisturbed. Be sure pH is adjusted with the correct CO_2 concentration.

Staining and Counting Procedure (nonsterile)

Pour off medium. Add 1 to 2 ml PBS. Swirl. Pour off. Repeat. Fix 5 min by adding 5 ml 95% ethanol. Pour off. Stain 20 to 30 min by adding 5 ml Giemsa stain. Pour off. Rinse with distilled H_2O. Air dry.

Count number of colonies by placing dish on counting page that is ruled in squares. Record colony count:

1.	3.	5.	7.	9.
2.	4.	6.	8.	10.

Average _____.

Table 15-2.   Single Cell Plating for Cloning

Culture No. _____   Date _____.

Select a bottle of cells that are nearly confluent.

Trypsinize by usual routine.

Add 10 ml of medium plus 15 to 20% serum. Suspend cells well with Pasteur pipette and perform cell count.

Counts per Large
    Corner Squares   8 squares   Divide by 8 $\times 10^4 =$
1 side of chamber   cells/ml $= n$

1   2   3   4

Other side of chamber
1   2   3   4

$$\frac{n}{1} = \frac{1 \times 10^6}{ml}$$

Use dispo 1-ml pipettes whenever possible, for cell transfers.

Use 10-ml dispo-plastic tubes.

Add _____ ml to make 10 ml suspension in medium plus serum (1×10^6 cells/10 ml).

Add 0.2 ml of cell suspension to 9.8 ml medium plus serum (2×10^4 cells/10 ml).

Add 1.0 ml of cell suspension to 9.0 ml medium plus serum (2×10^3 cells/10 ml) (200 cells/ml).

Be sure cells are well suspended before each dilution.

Add 10 ml medium plus serum to each 100×20-mm plastic tissue culture Petri dish.

To _____ Petri dishes labeled 100 cells, add 0.5 ml of cell suspension and agitate gently.

To _____ Petri dishes labeled 50 cells, add 0.25 ml of cell suspension.

To _____ Petri dishes labeled 40 cells, add 0.2 ml of cell suspension.
   (optional)

To _____ Petri dishes labeled 20 cells, add 0.1 ml of cell suspension.

Place dishes (level) in humidified desiccator or incubator at 37°C.

Adjust pH with the proper CO_2 concentration.

Dishes should not be disturbed for about 3 days.

trypsinization time at this point may prevent the cells from growing after subculture.

6. A 2-oz or 4-oz prescription bottle may be used for subculture. Add 10 ml of medium plus at least 15% serum (20% may be used). With a Pasteur pipette, add just enough medium from the 2-oz or 4-oz bottle to fill the cloning ring (about 0.5 ml). Place the end of the pipette lightly against the bottom of the center hole in the ring and rinse the medium up and down several times. Check carefully to make sure the ring is not leaking. If it is, the clone will need to be discarded. Add medium and cells to the subculture bottle. Repeat the rinsing of cells from inside the ring once or twice more. Check under the inverted microscope to make sure the ring is empty.

7. Incubate the bottle at 37°C with the proper pH adjustment (10% CO_2 for D & V).

8. At least 1 re-cloning is recommended for established cell lines. Human cells are difficult to re-clone.

Reference: Puck, T. T., Marcus, P. I., and Cieciura, S. J.: Clonal Growth of Mammalian Cells *in Vitro*. Growth Characteristics of Colonies from Single HeLa Cells with and without a "Feeder" Layer. J. Exp. Med., *103:*273, 1956.

Capillary Technique

Reference: Sanford, K. K., Covalesky, A. B., Dupree, L. T., and Earle, W. R.: Cloning of Mammalian Cells by a Simplified Capillary Technique. Exp. Cell Res., *23:*361, 1961.

Coverslip Method for Selecting Single Cells

Reference: Martin, G. M., and Tuan, A.: A Definitive Cloning Technique for Human Fibroblast Cultures. Proc. Soc. Biol. Med., *123:* 138, 1966.

REPLICATE PLATING PROTOCOL

1. According to routine trypsinization procedure prepare single cell suspension:
   Add 5 to 10 ml balanced salt solution (PBS), pour off, add PBS, pour off.

Add 0.25% trypsin, 1 ml to 8-oz bottle or 1.5 ml to 250-ml plastic flask.

Incubate till cells come off. Add 9 ml of medium to 8-oz bottle or 8.5 ml to 250-ml plastic flask, suspend cells well, and perform cell count. With Pasteur pipette, inoculate both sides of a hemocytometer. Count 4 large corner squares on each side of the chamber. Total the 8, divide by 8, $\times 10^4 =$ total number of cells/ml. If more than one bottle of cells is used, be sure to pool cell suspension from each bottle before performing cell count. (Multiply by number of ml to obtain total number of cells.) As a general rule, one confluent 8-oz prescription bottle will contain about 2×10^6 human diploid cells. A 250-ml plastic tissue culture flask (Falcon Plastics) will contain more cells (up to 4×10^6).

2. Calculate total number of cells required for the entire experiment from the number of dishes or bottles needed and the number of cells to be placed in each.

3. Suspend cells in the total amount of medium required for the experiment, calculated from the number of dishes or bottles needed and the amount of medium to be placed in each.

4. Dispense aliquots of cell suspension to each dish or bottle. Work quickly and keep cells well suspended by frequent pipetting. Place containers in 37°C incubator at the proper CO_2 concentration as soon as possible to restore pH and temperature.

Method of: Priest, J. H.: Unpublished.

"DEAD END" CELL COUNTS ON PETRI DISHES AND BOTTLES

"Dead end" cell counts on 60-mm Petri dishes and 8-oz prescription bottles are not recommended if cells are to be subcultured from these containers.

1. Decant medium, shaking out last drops.
2. Add balanced salt solution (PBS), rinse, pour off. Repeat.
3. Aspirate all remaining liquid with Pasteur pipette.
4. Add 1 ml 0.25% trypsin-0.25% Versene (T-V solution; see Chapter 12) to 8-oz bottle or 0.5 ml to 60-mm Petri dish.

5. Incubate till cells come off (less than 5 min).
6. Add 9 ml PBS to 8-oz bottle or 9.5 ml to Petri dish and suspend cells well with Pasteur pipette.
7. Inoculate 2 slides of a hemocytometer.
8. Count 5 large squares on each side of chamber.
9. Total the 10 counts, $\times 10^4 =$ total number of cells in 10 ml or per culture.
10. Adjust volumes for different size culture containers.

VIABLE CELL COUNT

1. Place 0.5 ml of cell suspension in a small tube.
2. Add 0.1 ml of 0.4% aqueous trypan blue and mix thoroughly.
3. Allow to stand 10 min (or incubate 5 min at 37°C).
4. With Pasteur pipette, fill a hemocytometer.
5. Count total cells and unstained cells.
6. Assuming those cells that have excluded dye to be viable, express the results as % of viable cells.

NEUTRAL RED VITAL STAIN

1. Add 0.5 ml neutral red vital stain (1:1300) per 5 ml of culture medium in culture container, and return to incubator for 10 to 15 min.
2. To obtain continued growth of the cells, remove the dye-containing medium and replace with fresh growth medium.

MITOTIC INDEX

1. A mitotic index may be determined from a chromosome preparation, provided metaphases are not broken.
2. Be sure slides are representative of all the cells in each culture container (preferably make up all the cells).
3. Examine a minimum of 1000 cells, from at least two slides.
4. Record number of metaphases per number of cells counted. Express as a percentage.

5. A routine should be used for choosing the portion of the slide to be counted. It is probably best to run the entire length of the slide an appropriate number of times, so that the entire slide is sampled at random.

CELL GROWTH RATE AND CELL CYCLE EVENTS

Simple and semiquantitative methods for estimating cell growth rate and cell cycle events (Fig. 15-1) will be described here.

Growth Rate—Estimation of Mean Cell Cycle Time from Population Doubling Time during Logarithmic Growth of Culture

1. Use replicate plating protocol and procedure for "dead end" cell counts (this chapter).
2. Select a nearly confluent bottle of cells to be studied.
3. Inoculate 3×10^5 cells in 12 to 15 ml of medium in each of 8 250-ml plastic tissue culture flasks (Falcon Plastics). (Petri

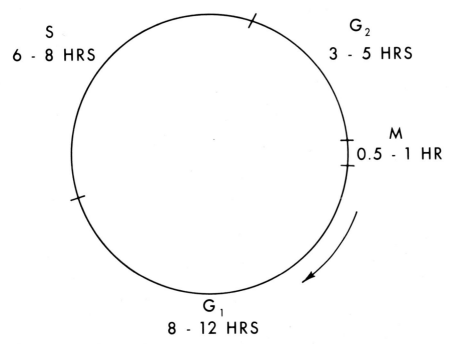

Figure 15-1. *Mammalian diploid cell life cycle.*

dishes or prescription bottles may also be used; the surface area should be large enough to allow for at least 2 days in log growth before saturation density is reached.) If fewer than 2×10^5 cells are inoculated initially, mammalian diploid cells will undergo a prolonged lag phase prior to log growth.

4. Incubate at 37°C with CO_2 adjusted for proper pH.

5. Perform cell counts on duplicate flasks or dishes at 24-hr intervals from the time of replicate plating for a total of 4 intervals. The entire analysis takes 5 days.

6. Plot cell counts (ordinate) against time in culture (abscissa) on semi-log paper. Including the initial number of cells at 0 time, 5 intervals are plotted.

7. *Interpretation:*
   The cells are in log growth when a straight line can be drawn through 3 plots. Determine the time (abscissa) for the cell number to double (ordinate) during log growth of the cells. This time approximates the mean population doubling time which in turn approximates the mean cell cycle time.

NOTE: The time between two mitoses can be determined by phase cinemicroscopy and represents the cell cycle time for one cell. The mean of a number of these determinations represents a mean cell cycle time. Measurements of this type indicate that the times for individual cells may differ widely from the mean (*i.e.,* there are fast and slow "dividers").

Length of Synthesis (S) Period of Cell Cycle—Estimation from Percent of Cells Incorporating Pulse Label of Tritiated Thymidine

1. Apply H_3TdR for 10 to 20 min to monolayer cells estimated to be in logarithmic growth in culture medium without thymidine. A final concentration of 2 to $3\mu c/ml$ of H_3TdR is recommended. Pour off the old medium and add medium containing the appropriate concentration of H_3TdR and no TdR, prewarmed to 37°C. Continue incubation for 10 to 20 min.

2. Termination of the label and the onset of processing of the cells for interphase autoradiography are both accomplished at the same time. Pour off medium and discard down the sink with running tap H_2O. Aspirate last drops and discard with running H_2O.

208

3. Add balanced salt solution (PBS, Chapter 12), rinse, pour off with running H_2O, aspirate last drops as above. Repeat this rinse.

4. Add 1 ml 0.25% trypsin-0.25% Versene in PBS (Chapter 12) and incubate 1-plus minutes until cells come off.

5. Add PBS and centrifuge in graduated centrifuge tube at 600 to 800 rpm for about 8 min. Discard supernatant with running H_2O and aspirate last drops.

6. Add to 0.5 mark with PBS. Add to 2.0 mark with distilled H_2O. Suspend cells gently. Incubate at 37°C or leave at room temperature for 10 min. This hypotonic step swells the interphase nuclei and makes them easier to study by autoradiography.

7. Shake gently and centrifuge as before. Aspirate all supernatant as close to the top of the button as possible.

8. Add 1 to 2 ml of 1:3 fixative (1 part glacial acetic acid to 3 parts absolute methanol) carefully down side of tube, without disturbing the button of cells.

9. Allow to stand at room temperature for 30 min.

10. Resuspend cells by pushing out with Pasteur pipette. Centrifuge as before.

11. Add about 0.2 ml 45% acetic acid to make a dense cell suspension and prepare air-dried slides (detailed in Chapter 7).

12. Rinse slides gently in running tap H_2O for at least 4 hr prior to the application of emulsion. Air dry. (See Chapter 9, sections on liquid emulsion and stripping film.)

13. Apply photographic emulsion, expose, and develop. (See Chapter 9.)

INTERPRETATION

1. Examine at least 200 random interphase cells. Record the total number of cells and the number labeled. Express results as percent of labeled cells.

2. From the estimation of the mean cell cycle time (T) and the percent of labeled cells (%), an initial *estimation* of the mean duration of synthesis (S) can be calculated: $S = T (\%/100)$.

Example: T = 20 hr
%= 30
$S = 20 \times 0.3 = 6$ hr

209

3. This estimation assumes that the total cell number is constant and that other restrictive conditions are also true—not normally the case in growing cultures. Various correction formulas have been developed (see the following reference by Cleaver).

Reference: Cleaver, J. E.: The Relationship Between the Duration of the S Phase and the Fraction of Cells which Incorporate H_3-thymidine During Exponential Growth. Exp. Cell Res., *39*:697, 1965.

Length of G_2 Period of Cell Cycle—Estimation from Time Necessary for Labeled Mitoses to Appear after a Pulse Label with Tritiated Thymidine

1. Apply H_3TdR for 10 to 20 min to 6 replicate cultures of monolayer cells estimated to be in logarithmic growth in culture medium without thymidine. (See section, this chapter, on replicate plating; and Chapter 9 on autoradiography.)
2. Prepare autoradiographs routinely as for chromosome analysis, using hypotonic treatment of the cells and air drying of slides. (See Chapter 5, page 66.) Be sure to discard radioactive solutions down the sink with running H_2O. Process 1 bottle at 2 hr after onset of label, 1 at 2½ hr, 1 at 3 hr, 1 at 3½ hr, 1 at 4 hr, 1 at 4½ hr.
3. Examine at least 50 metaphases from each interval. Record the total number of metaphases and the number labeled. Express results as percent of labeled metaphases.
4. The interval when 50 percent of the metaphases are labeled represents the mean length of G_2.

Length of G_1—Indirect Estimation when Lengths of S, G_2, and Mean Cell Cycle Time Are Known

1. The length of mitosis (M) for mammalian cells under the usual conditions of asynchronous log growth monolayer culture is less than 1 hr, as determined by phase cinemicroscopy.
2. Therefore: mean total cell cycle time, less the lengths of S, G_2, and M, gives an estimate of G_1.

SYNCHRONIZATION OF MAMMALIAN CELLS

Many procedures are available for the synchronization or phasing of mammalian cells to make a cell population more uniform in progress through the cell cycle or portions of it. For instance, for an individual interested in examining chromosomes, it is very useful to sample the cells at a time when the majority are in mitosis.

This section includes a discussion of phasing procedures that are applicable to human diploid cells in serial monolayer culture. Striking characteristics of the synchronization thus achieved are that it is short-lived and never reaches 100%. These facts are not surprising when one recalls the wide range of behavior of each cell within the whole population. Any procedure applied to an entire population does not take into account the individual states of the cells. Furthermore, cells starting a portion of the cell cycle together will not stay together if they are moving through the cycle at different individual rates.

References: Cameron, I. L., and Padilla, G. M., eds.: *Cell Synchrony. Studies in Biosynthetic Regulation.* New York, Academic Press, 1966.

Zeuthen, E., ed.: *Synchrony in Cell Division.* New York, Interscience Publishers, 1964.

Synchronization at Beginning of S—Effect of an Antimetabolite (FUdR) Which Stops DNA Synthesis and Accumulates Cells at End of G_1

1. Inoculate replicate cultures into culture medium without thymidine and maintain until exponential phase of growth.

2. Add FUdR (5-fluoro-2'-deoxyuridine, Hoffman-LaRoche, Inc., Nutley, N. J., Research Division) to a final concentration of 0.1 μgm/ml (4×10^{-7} M). Also add uridine 6×10^{-6} M. (This synchronization procedure is also described in Chapter 9 on autoradiography, page 123.)

3. After 16 to 20 hr of incubation at 37°C (one mean cell cycle, or slightly less), relieve the block to DNA synthesis by pouring off the medium containing FUdR and adding medium containing 6×10^{-6} M thymidine (TdR).

4. Sample the cells at intervals after reversal of the FUdR block with exogenous TdR. Reversal of the block is timed as the beginning of S. Synchrony is lost before one cell cycle is completed. A maximum of 75% of human diploid cells are in S immediately following reversal of the block, as compared to an asynchronous level of 30%.

Reference: Priest, J. H., Heady, J. E., and Priest, R. E.: Delayed Onset of Replication of Human X Chromosomes. J. Cell Biol., *35:* 483, 1967.

211

Synchronization at Beginning of S—Effect of Excess Thymidine (TdR) Which Stops DNA Synthesis and Accumulates Cells in G_1 or Early S

A double block is described here. No agent that stops all S cells can produce a strongly synchronized culture. When a single block is relieved, these cells will start up from where they are in S. However, if a second block is applied when all the cells are out of S, relief of this block should place a high percent of the cells at the beginning of S.

1. To cells in logarithmic growth phase, add 2.5 mM TdR for 24 hr.
2. Wash 3 times in balanced salt solution.
3. Incubate in normal medium for 15 hr (or a period long enough for all cells to finish S).
4. Apply TdR a second time for 24 hr.
5. Wash the cells and return to normal medium or process as for experimental outline.

References: Kasten, F. H., Strasser, F. F., and Turner, M.: Nucleolar and Cytoplasmic Ribonucleic Acid Inhibition by Excess Thymidine. Nature, *207:*161, 1965.

Puck, T. T.: Studies of the Life Cycle of Mammalian Cells. Cold Spring Harbor Symposium, *29:*167, 1964.

Synchronization during Mitosis—the Metaphase Shake Method

Mitoses in a monolayer culture are less firmly attached to the surface of the culture container than are interphase cells. The metaphase shake method is designed to shake off mitoses into the overlying medium (calcium-free) and leave interphase cells in the attached monolayer. Unfortunately, human diploid cells in mitosis are almost as firmly attached as interphases, and the method is less successful for human cells, but is sometimes useful.

1. About 6 hr before mitoses are desired, pour off culture medium, rinse twice, and replace with calcium-free medium. Re-incubate.
2. Shake and rotate bottle to free the mitoses into the overlying medium.
3. Centrifuge overlying medium at 800 rpm for 5 min and resuspend the pellet in complete medium or as needed for the experimental outline.

Reference: Robbins, E., and Marcus, P. I.: Mitotically Synchronized Mammalian Cells: A Simple Method for Obtaining Large Populations. Science, *144:*1152, 1964.

MODIFICATION

Mitoses are accumulated in the presence of Colcemid and are separated with a brief trypsin treatment.

1. Add 0.06 μgm/ml Colcemid (CIBA) for no longer than 4 hr to monolayer cells that have been subcultured about 18 hr previously. It is suggested that the cells used for this type of synchronization should be subcultured at 24-hr intervals or, depending on the rate of cell division, at intervals that prevent the monolayers from reaching confluency or stationary phase.
2. Remove medium and add 5 ml 0.1% trypsin (cold) for 45 sec. During trypsinization, shake the cultures gently horizontally.
3. Drain off trypsin to bottom of container. Transfer suspended mitotic cells to centrifuge tube and centrifuge at 600 to 800 rpm for 2 min or more until a button forms.
4. Resuspend in fresh medium and proceed as needed for the experimental outline.

Reference: Stubblefield, E., Klevecz, R., and Deaven, L.: Synchronized Mammalian Cell Cultures. I. J. Cell. Physiol., *69:*345, 1967.

Not Recommended for Diploid Mammalian Cells in Monolayer Culture, Particularly Human

1. Temperature shock.
2. Some types of double blocks such as:

   a. Double FUdR
   b. Excess thymidine plus FUdR
   c. Cold shock plus FUdR

Reference: Priest, J. H.: Personal observations.

Glossary

The italicized words in Chapters 1 and 2 and in the introductions to the other chapters, are defined in this section. Cytogenetic nomenclature is also discussed in Chapter 3, and tissue culture nomenclature in Chapter 13.

Aneuploid—deviation from the basic number or from exact multiples of the basic number in a chromosome series. *(Aneuploidy)*

Antigenic—having the ability to stimulate production of an antibody, or to stimulate an allergic reaction.

Arm (chromosome)—the portion of a chromatid on one side of the centromere of a metaphase chromosome.

Arm ratio—the length of the longer arm of a metaphase chromosome relative to the shorter arm.

Association (chromosome)—proximity of certain portions of different chromosomes. *(Associated)*

Autoimmunity—a situation in which an individual produces antibodies (or an allergic reaction) against his own tissue proteins.

214

Autoradiography—from *auto*, meaning self, and *radiography*, meaning the practice of producing a picture upon a sensitive surface, by some form of radiation, usually other than light. (*Autoradiograph*)

Autosomes—non-sex chromosomes (22 pairs in humans). (*Autosomal*)

"Balanced" structural rearrangement (chromosome)—rearrangement of chromosome material, without genetic effect.

Bivalent—two paired homologous chromosomes during meiosis.

"Blastoid" reaction—the process by which a small lymphocyte is transformed *in vitro* into a large, morphologically primitive, "blast-like" cell capable of undergoing mitosis.

Break (chromosome)—interruption in staining of chromosome arm, with disturbance of alignment of the portions on either side of the interruption.

Cell cycle (cell life cycle)—the cycle in the life of a cell which includes mitosis (M) and interphase. The cell progresses from M to G_1, to S, to G_2, and on to M again.

215

Centromere (kinetochore, primary constriction)—a nonstaining area on a chromosome, separating the chromosome arms; the point of attachment of the chromosome to the mitotic spindle. (*Centromeric*)

Centromere index—the ratio of the length of the shorter arm to the length of the whole chromosome.

Chromatid—one of two structurally distinguishable longitudinal subunits (by light microscopy) of a metaphase chromosome.

Chromatin—areas of a cell nucleus that stain in some manner with a DNA stain. (See also *sex chromatin*.)

Chromomere—"beadlike" substructure on meiotic prophase chromosome, constant in size and position.

Cinemicroscopy—motion picture microscopy.

Daughter cell—one of the two products of mitosis.

Deletion (chromosome)—absence of part of a chromosome.

Diploid—the double basic number in a chromosome series (2x). (*Diploidy*)

DNA—deoxyribonucleic acid; a constituent of chromosomes.

Duplication (chromosome)—repetition of part of a chromosome.

Endoreduplication (endopolyploidy)—the result of a partial mitosis involving doubling of chromosome number without division of nucleus or cytoplasm.

Euchromatic—relating to areas of the nucleus that do not stain "differently" with DNA stains.

Exchange (chromosome)—transfer, morphologically, of chromosome material between chromosomes.

Exchange figure (chromosome)—partial pairing of chromosomes at mitosis.

Fibroblast-like—spindle-shaped appearance, used to describe a cell that resembles a fibroblast but has an unknown derivation or potentiality.

Fragment (chromosome)—a portion of a chromosome.

Gamete—mature male or female germ cell.

Gap (chromosome)—interruption in staining of chromosome arm, without disturbance of alignment of the portions on either side of the interruption.

216

Genotype—the genetic constitution. (*Genotypic*)

Haploid—the basic number in a chromosome series (x). (*Haploidy*)

Hermaphrodite—both male and female.

Heterochromatin—nuclear areas that stain "differently" with DNA stains. (*Heterochromatization; Heterochromatic*)

Heterokaryon—cell with two or more genomes of different types.

Homologous (chromosomes, genes)— similar.

Isochromosome—a chromosome consisting of identical arms on either side of the centromere.

Karyotype—a systematized array of the chromosomes of a single cell prepared either by drawing or photography, with the extension in meaning that the chromosomes of a single cell can typify the chromosomes of an individual or even a species. (*Karyotyping*)

Length (chromosome)—the total distance between the opposite ends of a metaphase chromosome.

Long-term culture—*in vitro* maintenance of cells in serial culture.

Map (chromosome)—a linear diagram of the location of genes on a chromosome. (*Mapped*)

Meiosis I, II—the 1st and 2nd cell divisions of meiosis.

Metaphase I, II— the 1st and 2nd meiotic metaphases.

Mitogenic—capable of inducing mitosis.

Mitotic index—the number of mitoses per total number of cells in a given sample, expressed as a percent.

Monolayer—a single layer of cells growing on a surface.

Monosomic—the double basic number in a chromosome series minus one; one pair contains 1 instead of 2 chromosomes. (*Monosomy*)

Mosaicism (chromosomal)—a situation with more than one chromosomal cell type. (*Mosaic*)

Nondiploid—a chromosome number other than the double basic number or diploid. (*Nondiploidy*)

Nucleolus—a nuclear structure that stains with both DNA and RNA stains.

Oöcyte—immature female germ cell.

Oögenesis—the formation of female germ cells.

Oögonium—primordial female germ cell. (*Oögonia*—plural)

217

Organ culture—the maintenance or growth of organ primordia or the whole or parts of an organ *in vitro* in a way that may allow differentiation and preservation of architecture or function.

Ovum—mature female germ cell. (*Ova*—plural)

Phenotype—external appearance or constitution of an individual. (*Phenotypic, Phenotypically*).

Polyploid—general designation for multiples of the basic chromosome number, higher than diploid which is twice the basic number. (*Polyploidy*)

Population doubling—refers to doubling of cell number in a population of cells, for example, when 1×10^6 cells increases to 2×10^6 cells.

Primary explant—an excised fragment of a tissue or an organ used to initiate an *in vitro* culture.

Pulverization (chromosome)—loss of metaphase chromosome morphology, with preservation of enough chromosome material to stain with a DNA stain.

Replication, DNA—the process by which new DNA is made from a DNA template.

Ring (chromosome)—attachment of opposite ends of a chromosome to form a ring (a deletion is implied in this situation).

RNA—ribonucleic acid.

Satellite—DNA staining structure on the distal end of a chromosome arm and separated from it by a secondary constriction. (*Satellited*)

Secondary constriction—a nonstaining area on a chromosome arm.

Segregation (chromosome)—the distribution of chromosomes to daughter cells.

Sex chromatin (Barr body)—a characteristic area of heterochromatin in the nucleus, composed of X chromosome material that stains heavily with a DNA stain.

Sex chromatin negative (chromatin negative)—the sex chromatin body is not present in a cell nucleus.

Sex chromatin positive (chromatin positive)—the sex chromatin body is present in a cell nucleus.

Sex chromosomes—XX in human female; XY in human male; 1 pair is normally present in each individual.

218

Short-term culture—*in vitro* maintenance of cells for limited periods of time, not involving subculture.

Somatic—pertaining to cells other than germ cells.

Spermatocyte—immature male germ cell.

Spermatogenesis—the formation of male germ cells.

Synthesis period (S)—the period during interphase when DNA is replicated.

Tetraploid—the quadruple basic number in a chromosome series (4x). (*Tetraploidy*)

Transcription, RNA—the process by which new RNA is made from a DNA template.

Transformation (cell)—change to a different function or appearance induced by the introduction of new genetic material.

Translocation (chromosome)—portion of a chromosome located on a chromosome other than the normal one.

Triploid—the triple basic number in a chromosome series (3x). (*Triploidy*)

Trisomic—the double basic number in a chromosome series, plus one; one pair contains 3 instead of 2 chromosomes. (*Trisomy*)

Tritiated thymidine—thymidine containing tritium, one of the isotopes of hydrogen, which is unstable and decays with the emission of a weak β-particle.

Watson-Crick double helix—a model for the molecular structure of DNA.

Zygote—the first product of the union of sperm and egg.

Index

Page numbers in *italics* refer to illustrations; page numbers followed by t refer to tables.

All entries refer to human chromosomes unless otherwise noted.

221

229